TRACTION ENGINE RALLIES

Cover image: The spectacle of heavy haulage at the Great Dorset Steam Fair as road locomotives combine to move some fifty tons of Portland stone on the low-loader trailer. In the lead is Fowler B6 10nhp engine No. 17212 *Wolverhampton Wanderer*, which was supplied new in 1929 to John Thompson, Ettingshall, Wolverhampton to deliver boilers to their customers.

Frontispiece: The one that started it all. The late Arthur Napper's 1902 Marshall *Old Timer*. This became a regular at the Knowl Hill rally where it is seen in 1992. The other engine in the historic races of 1950, Miles Chetwynd-Stapylton's Aveling & Porter *Ladygrove* was later sold on and subsequently scrapped.

TRACTION ENGINE RALLIES

AN APPRECIATION OVER SEVENTY YEARS 1950–2019

Malcolm Batten

PEN & SWORD TRANSPORT

AN IMPRINT OF PEN & SWORD BOOKS LTD.
YORKSHIRE – PHILADELPHIA

First published in Great Britain in 2023 by
Pen and Sword Transport
An imprint of
Pen & Sword Books Ltd
Yorkshire - Philadelphia

Copyright © Malcolm Batten, 2023

ISBN 978 1 39908 167 2

The right of Malcolm Batten to be identified as Author of this work has been asserted by him in accordance with the Copyright, Designs and Patents Act 1988.

A CIP catalogue record for this book is available from the British Library.

All rights reserved. No part of this book may be reproduced or transmitted in any form or by any means, electronic or mechanical including photocopying, recording or by any information storage and retrieval system, without permission from the Publisher in writing.

Typeset in 11/14 Palatino
Typeset by SJmagic DESIGN SERVICES, India.

Printed and bound in India by Replika Press Pvt. Ltd.

Pen & Sword Books Ltd incorporates the Imprints of Pen & Sword Books Archaeology, Atlas, Aviation, Battleground, Discovery, Family History, History, Maritime, Military, Naval, Politics, Railways, Select, Transport, True Crime, Fiction, Frontline Books, Leo Cooper, Praetorian Press, Seaforth Publishing, Wharncliffe and White Owl.

For a complete list of Pen & Sword titles please contact

PEN & SWORD BOOKS LIMITED
47 Church Street, Barnsley, South Yorkshire, S70 2AS, England
E-mail: enquiries@pen-and-sword.co.uk
Website: www.pen-and-sword.co.uk

or

PEN AND SWORD BOOKS
1950 Lawrence Rd, Havertown, PA 19083, USA
E-mail: Uspen-and-sword@casematepublishers.com
Website: www.penandswordbooks.com

CONTENTS

Introduction	6
Early days 1950s	11
1960s flashback	18
'One-off' rallies	22
Grand Reunion of Burrell's Steam Engines 1972	22
Clapham Common, London 1973	25
Clapham Common, London 1977	27
Northampton 1996	29
Lincoln 2000	30
Fawley Hill 2014 & 2016	33
Annual rallies that have ceased	39
Birmingham Museum of Science & Industry c.1969–97	39
Knowl Hill Steam Rally 1970–2004	46
Lambeth Country Show c.1973–1990s	53
Expo Steam, Peterborough Showground 1972–89	56
Rushmoor Arena 1975, 1978–98	62
Essex Steam & Country Show / Essex Country Show 1986–2016	70
Preston Rally & Country Fair 2003–14	79
Annual rallies ongoing	88
HCVS London to Brighton Run 1962–date	88
Banbury Vintage Vehicle and Country Fayre 1969–date	98
Bedfordshire Steam & Country Fayre 1957–date	104
Chiltern Steam Rally 1962–date	126
Hertfordshire Rally 1964–date	134
Island Steam Show 1975–date	139
Medway Festival of Steam & Transport, Chatham Historic Dockyard 2002–date	145
Netley Marsh Steam & Craft Show 1971–date	152
Weeting Steam Engine Rally 1968–date	160
Great Dorset Steam Fair 1969–date	173
Bibliography	192

INTRODUCTION

In 1969 Barry J. Finch wrote a book entitled *A Rally of Traction Engines* and this seems a very appropriate collective noun for these vehicles. Their commercial life largely came to an end in the 1950s and early 1960s. It was also at this time that preservation and display in the form of rallies came into being. While a few engines were kept by their commercial owners out of sentiment and some others were acquired by local museums, there was no concerted effort to save a representative collection – it was just whatever was saved by individuals. However, one organisation, the Road Locomotive Society, was formed in 1937 whose stated aim is 'The enhancement of education and research into the past history of all types of self-propelling steam engines (other than those running on rails)'. They were not specifically involved in preservation, although they did acquire an early Aveling & Porter traction engine (No. 721 of 1871) and presented it to the Science Museum in 1951.

It is generally acknowledged that the traction engine rally scene has its origins in a race between two engine-owning farmers from Appleford, Berkshire (now in Oxfordshire) on 30 July 1950. Arthur Napper and Miles Chetwynd-Stapylton raced their engines over a one-mile course, up and down a field at Bridge Farm. The prize was to be a firkin of ale (nine gallons). The winner was Arthur Napper in his Marshall engine BH 7373 *Old Timer* after the rival engine, Aveling & Porter No. 8923 *Ladygrove* had forged ahead but then suffered from leaking tubes. An audience of twenty-five turned out to witness the event. A re-run was then held on 20 August. As this had been publicised in the *Didcot Advertiser* a crowd of over 250 gathered to see Arthur Napper's Marshall again win the race. This was also reported on in the *Daily Mirror*. A new challenge was then issued by Giles 'Doc' Romanes from Bray. A third race took place at Nettlebed on 21 June 1951 when Arthur Napper once again won, beating Giles' 1919 Wallis & Steevens engine No. 7683 *Eileen*. The event had been well publicised and over a thousand people turned out to see this. A collection was held at the race, which raised some £80 for a local hospital, thus demonstrating the potential charitable benefits of such events.

A first traction engine rally was held at Bridge Farm on 8 June 1952 when eight engines turned up to exhibit and in some cases race against each other. The last Appleford Rally took place in 1971, but the seed had been truly sown.

The rally movement soon grew as area preservation societies were formed. The first of these was the East Anglian Traction Engine Club (now Society) in 1955. They held their first rally at Saling Aerodrome, near Braintree, Essex on 9 July 1955. There were twenty-three engines present and between 5,000 and 6,000 visitors. By this time racing of engines had ceased in the interest of safety, although there was an Obstacle Steering Competition. One interesting competition that took place was Setting to Drum where an engine had to line up with a threshing machine, fit the driving belt and run the machine up to working speed - all while being timed against the clock. Many of the drivers and crew at this time would have had previous working experience with steam. All of the engines that attended remain in preservation today, many of them still based in East Anglia. Also established in 1955 was the West of England Steam Engine Society.

A first rally in Norfolk took place at Woodton Old Hall in 1956 with just six engines. But the rally grew in subsequent years and in 1959 the Norfolk Steam Engine Club was formed and took over the organisation of the event. The Woodton rally continued until 1968 after which the NSEC rally moved to Strumpshaw Hall.

The Strumpshaw rally has continued to be held annually, but since 1993 has no longer been organised by the NSEC.

A significant 1950s rally was that organised on 5 August 1957 by John Crawley at Woburn Abbey. This was the first rally held by the Bedford Steam Engine Preservation Society, and forty-eight engines attended. The site would host further rallies, and in 1962 would be the setting for the feature film *The Iron Maiden* starring John's Fowler showman's road locomotive No. 15657 *Kitchener*, a film that really brought the public's attention to the rally scene.

The first rally to be held in Lincolnshire was at North Hykeham in 1959. Organised by the Lincolnshire Steam Engine Preservation Society, formed in 1959, from 1960 this became a two-day event with proceeds going to the Hykeham Memorial and Playing Fields Association. The Society is still going strong, but their rally now takes place at the Lincolnshire Showground.

A notable rally for its time was the first 'Raynham Day' at Raynham Hall in North Norfolk on 16 September 1962. This was organised by Dick Joice, a Norfolk farmer and engine owner. A massive total (for that time) of fifty-one engines attended, eleven of which came from the late George Cushing of nearby Thursford. Many of these are now part of the Thursford Collection.

Things were rather slower north of the border. The Scottish Traction Engine Society was formed in 1961 but the first rally did not take place until 1968 in the grounds of the now demolished Hamilton Palace, continuing annually there until 1973.

Some of the early societies and rallies continued to flourish and a number of these have now celebrated fifty or more years of activity, albeit not always on the same site throughout. For instance, the sixtieth anniversary of the first Appleford race was celebrated at the nearby Woodcote Rally in 2010. The Woodcote Rally itself dates back to 1964 and celebrated its 50 years in 2013.

Other rallies flourished for a while but then ceased for varying reasons, and there have also been a number of 'one-off' events.

These gatherings or rallies can take place in a variety of locations. Traditionally they have been on farm sites where fields have been made available. This is quite appropriate, given that many engines were used in agriculture. But rallies also take place in such sites as parks, the grounds of agricultural colleges and on county showgrounds. They can vary in size from small local events with just a handful of exhibits to the Great Dorset Steam Fair, the world's largest such heritage show with more than 200 traction engines out of a total of over 2,000 exhibits. Many of the heritage railways and open-air museum sites now incorporate a traction engine event as part of their annual programme. Thus, the Great Central Railway regularly has a steam fair over the Easter weekend. The North of England Open Air Museum at Beamish has staged some spectacular events in recent years. Then there are also road runs, either local affairs organised by a steam club or nationally advertised events for commercial vehicles of all types. Details of the hundreds of events held throughout the year are published in guides produced by the two main magazines that have been published for enthusiasts of the traction engine and general heritage preservation scene – namely *Old Glory* and *Vintage Spirit*.

The early rallies were mainly ring events, where the entrants paraded around the ring while a commentary was given about each of the engines (and similarly for cars, tractors etc when it was their turn). When all the engines were assembled in the ring, there would be a mass sounding of their whistles followed perhaps by an opportunity for the public to come into the ring and take photos for a while before the engines dispersed. Before and/or after the main ring parade there would a few engines in the ring for members of the public to try their hand at supervised steering. There might be games like an obstacle race or a 'slow race' for engines or 'musical chairs' where children would ride on an engine while the music played then rush to claim a place at an oil drum or straw bale when it stopped. There might also be judging of the engines with cups or medals for best in class.

The initial concept of rallies has developed over the years. Instead of just ring events many now incorporate working areas where the different types of engines can be demonstrated doing the tasks for which they were built. Thus, there may be demonstrations of threshing, wood sawing, stone crushing etc, with the engines driving the machinery by belts from their flywheels. There may be road-making demonstrations where the rollers can strut their stuff. Showman's road locomotives and tractors get to power traditional rides and organs. At some sites, where fields are available, steam ploughing may be demonstrated. Finally, there is the magnificent sight of heavy haulage with road locomotives working in multiple, and here the Great Dorset Steam Fair with its 'play pen' and 'Watford Gap' hill reigns supreme.

These rallies do not just exhibit full size traction engines and their miniature cousins. They have also developed to display vintage tractors, cars, commercial vehicles, buses etc. There will be a fair, market, various forms of entertainment from folk music to sheep-shearing displays, numerous organs, sometimes accompanied by a stage show, model and craft tents, and of course a variety of food and drink outlets. In all, something for everyone of all ages.

For these rallies and other events to happen, we are indebted to those who had the foresight to save so many engines from the scrapyard, then to spends years and fortunes restoring them to their former splendour. But we also owe a great deal of thanks to those club committee members and friends who organise the rallies and give up their time to be marshals, stewards, judges and commentators at the events. Without these often-unsung heroes none of this would have happened.

In the 1950s, engines could be obtained for little more than their scrap value – sometimes less than £25. But with the costs of ongoing restoration and the finite number of surviving engines, prices are somewhat different now. In October 2020, Foden B6 showman's road locomotive No. 19782 *The Lion*, right at the top end of the market, came up for auction at Bonhams in London following the death of Arthur Thomson its original owner in preservation and shortly after, that of his brother who had inherited it. It was eventually sold for £800,000 to John Saunders for the Saunders collection. Interestingly, he had tried to buy *The Lion* before, but had been outbid by Arthur Thomson at the time.

The number of engines attending rallies has also increased over the years. This has been due to more of the existing engines initially saved being restored, some engines originally placed in playgrounds on withdrawal being later rescued and restored, engines originally built for export being repatriated from around the world, as well as examples of foreign-built engines being imported and finally the creation of 'new-build' engines built to original designs. There are still many engines that were initially put aside for eventual restoration that have yet to be restored. In 2008, a collection of unrestored engines that had been put aside in a garage at East Preston, near Littlehampton by the late John Lillywhite in the 1950s were 'rediscovered' and sold off to new owners.

As well as engines being imported and repatriated (although others have been exported), there have also been a number of visits by foreign based engines to British rallies. Some engines from Ireland are usually to be found at the Great Dorset Steam Fair, but others have visited from Holland where there is a significant preservation movement. However, the highlight has been the visit of several engines from New Zealand, which has probably the largest following for traction engines outside the UK. The Hawkins family brought Burrell No. 3130 of 1909 over from 1996-8. Then, in 2000, five Burrell engines were shipped over to make a guest appearance at the Great Dorset Steam Fair. They also attended other rallies including Weeting in 2001 thus giving them an opportunity to return to nearby Thetford where they were built. An outbreak of foot and mouth disease led to them not being allowed home at the end of the season and so they stayed on for another year. In 2010, five McLaren engines came over for a themed display at the GDSF and also attended other events. Since then, other engines have visited from New Zealand and elsewhere and some have made return visits.

Some British engines and their owners have also travelled to events in Europe and beyond since at least 1972.

The *Traction Engine Register* was first published in 1966. Compiled by John True and Brian Kinsey it was a typewritten and duplicated 28-page listing of nearly 1,100 road engines, portables and steam fire engines with a location, owner's name and brief history which sold for 2/6d. A second, printed edition came out in 1968, published by the Southern Counties Historic Vehicles Preservation Trust. This listed just over 1,900 engines. Now updated every four years, the twelfth edition published in 2016 and edited by Brian Johnson listed 3,826 engines in the UK and Ireland and now also included fairground centre and organ engines. Much of this increase was due to the number of imported engines. The thirteenth edition, 2020 lists 3,859 engines from 128 different manufacturers. Since 1978, the Register includes an appendix listing the road registration marks and the engine each relates to. This is the standard reference source and is invaluable to the enthusiast.

The umbrella organisation for the preservation movement is the National Traction Engine Trust (NTET), originally the National Traction Engine Club. This was set up in 1954 to ensure the preservation of road-going steam engines. Their membership comprises individual members and affiliated clubs from throughout the world. They produce a Code of Practice under which approved rallies are held. They set safety standards and provide technical advice, and campaign against any legislation that might adversely affect the preservation hobby. They also secure insurance and other related services for members and rally organisers; support the maintenance of records of engine manufacturers; and provide facilities to teach basic skills in engine management including attracting junior members through the Steam Apprentice Club, founded in 1979. An annual list of NTET approved rallies is produced – in the 2018 edition there were 62 such events in the United Kingdom and Ireland. A quarterly magazine *Steaming* has been published for members since 1957. There were over 3,500 members worldwide by 2016.

Other groups like the Steam Plough Club and Road Roller Association have been set up to preserve the history and keep the traditions alive for owners of these particular kinds of engines. Some of the road locomotives became worked under the joint banner 'Amalgamated Heavy Haulage' from the mid-1980s.

Some engines were modified during their working lives to meet the requirements of their owners. However, many more have been reconstructed in preservation and this has remained a controversial topic. In particular, several tractors have had showman's fittings added, while more radically numerous rollers have been rebuilt as tractors or traction engines. The NTET has taken a strong view on such 'conversions' while recognising the right of owners to do as they please with their own property. They issued a statement 'One of the major objectives of the NTET is to preserve our steam heritage, including the original design of our road steam vehicles, their use and maintenance. Therefore, the Trust does not support in any way the conversion of any steam road vehicle into a form which that particular machine never had during its working life.'

In this book I intend to feature a number of rallies, starting with some examples from the early years of the 1950s and 1960s. Then some 'one-off' events, a look at some of the rallies that no longer take place and finishing with examples of those that are still flourishing. I will endeavour to show something of the individual character of each rally, and some of the highlights of events that I have visited over the last fifty years. I was first taken to a rally at Ingatestone, Essex as a teenager in September 1966, and first visited a rally with a camera in 1972. The selection ends with 2019 as most events were cancelled in 2020-21 due to the Coronavirus pandemic and some have also been cancelled for 2022 because of uncertainties as to whether there would be another lockdown.

As I live in London, most of the locations featured are in the south of England. This doesn't mean that the rallies in the north, or in Scotland or Wales are less significant, just that I have only visited such events occasionally rather than regularly.

All photographs are by the author except where credited.

EARLY DAYS – 1950s

Many engines now in preservation were still working for their living in the early 1950s. Some steam wagons finished life as tar boilers, the steam keeping the tar hot for pouring. This is Sentinel DG6 No. 8562, FD 6603 of 1931 which was with Thames Tar Products and Contractors Ltd at their Beddington Lane Depot near Croydon on 2 December 1950. *John Meredith/Online Transport Archive*

Some steam rollers remained in commercial use into the early 1960s but that did not guarantee their preservation. Aveling & Porter No. 4855, ME 3266 was No. 5 in the fleet of Rowley Plant Hire Co. Ltd. when seen in 1950 but did not survive. The Marshall roller in the background No. 78724 OPC 765 was preserved but has since been converted to a tractor in 1975 – a fate that has befallen many rollers. *John Meredith/Online Transport Archive*

Above: Dr Romanes's Wallis & Steevens engine *Eileen* which lost the 1951 race to Arthur Napper's *Old Timer* is seen at the 1953 Appleford Rally. *John Meredith/Online Transport Archive*

Right: Tasker No. 1776, AP 9281 of 1918 at the Andover Rally held at Finkley Manor Farm on 12 May 1956. This engine is now one of a number of Tasker engines owned by Hampshire County Council and is in store at Winchester, not currently on public display. *John Meredith/Online Transport Archive*

Left: The oldest surviving Allchin engine, No. 669 of 1890 is seen at the West Sussex Traction Engine Rally at North Heath, Pulborough on 12 August 1956. Note the period fashion sported by the boy standing by the back wheel. *John Meredith/Online Transport Archive*

Below: At the Saffron Walden Steam Festival, 15 June 1957, Ransomes, Sims & Jefferies No. 41046 of 1930 takes part in the ring parade. This engine, registered VX 7317 is now resident in the Republic of Ireland. *John Meredith/Online Transport Archive*

Right: At the same event, an engine still very much with us and still in East Anglia is Ruston & Hornsby 6nhp engine No. 113043, CE 7977 *Oliver*. *John Meredith/Online Transport Archive*

Below: At the first Woburn Park Traction Engine Rally of 5 August 1957 is John Crawley's famous Fowler showman's road locomotive FX 6661 *Kitchener*. It was of course this engine that was to star in the 1962 film *The Iron Maiden*, the climax of which was filmed here at Woburn. The engine now carries both names and now forms part of the Scarborough Fair Collection. *John Meredith/Online Transport Archive*

Again at Woburn Park, Burrell No. 3847, CL4483 is belted to a threshing machine. At the time this was owned by Mr J. Bennie from Northamptonshire. Nowadays this has been rebuilt to its original format as a showman's road locomotive named *Princess Marina* and has been with the Searle family at Horsham since 2000. *John Meredith/Online Transport Archive*

Also at Woburn Park in 1957 is Foster 4 nominal horsepower (nhp) showman's tractor No. 14066 of 1915. Shown here with registration number PP 5649, this now carries the registration FE 1589. Note the period cars which would grace any modern rally – is that a Humber Super Snipe on the right?
John Meredith/Online Transport Archive

1960s FLASHBACK

Above: A rally at Chartridge, Buckinghamshire in the early 1960s, exact date uncertain but possibly 1963 or 1964. This rally was a precursor of the Chiltern Steam Rally now held at Prestwood, Bucks. Wallis & Steevens Roller HO 5834 of 1903 is parading around the ring. This has since been converted to tractor form and is now named *Little Olga*. No double ring or orange plastic fencing then. Those cars in the field would all be entrants in the historic car section at any modern rally! Reg Batten

Opposite above: At the same event is Aveling & Porter roller No. 9008, BH 7167 *George*. This spent its working life fairly locally as it was owned by Buckinghamshire County Council. The 2016 edition of *The Traction Engine Register* shows this engine as being based in Cornwall by then. Reg Batten

Opposite below: Also present was Burrell 'Gold Medal' tractor No. 3554, AH 0181 *King George V*. This engine has been owned by the Davis family of Prestwood since 1956 and is a regular attender at the Chiltern Steam Rally. It also attended a number of other early rallies including at Appleford and the first Woodcote Rally in 1964. Note the replacement front wheels. They were actually fitted during the engine's commercial life with a timber company from Swindon. The boy riding on the back is possibly taking part in a game of 'musical chairs' and will need to run for one of the oil drums when the music stops. Reg Batten

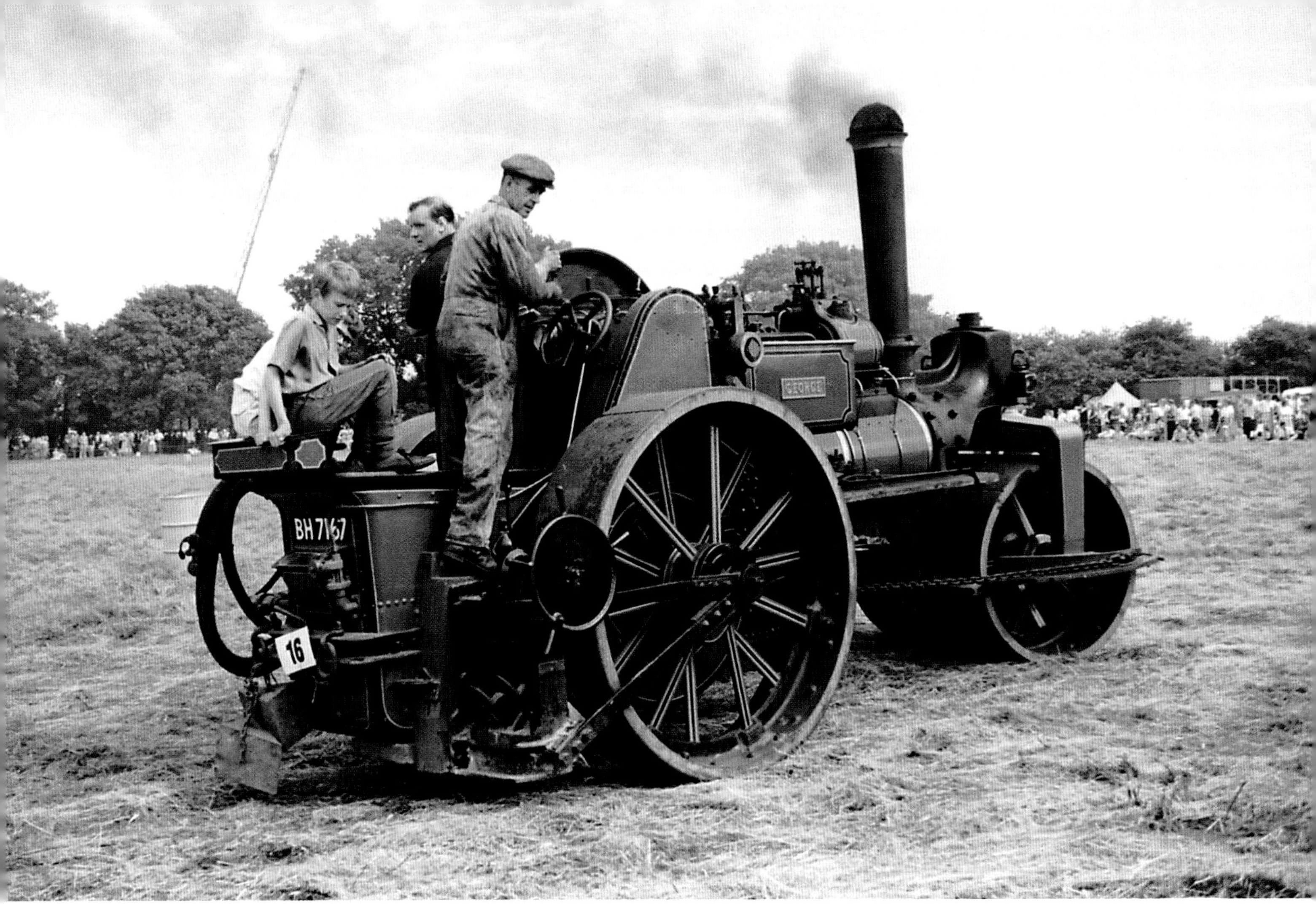
GEORGE
BH 7167
16

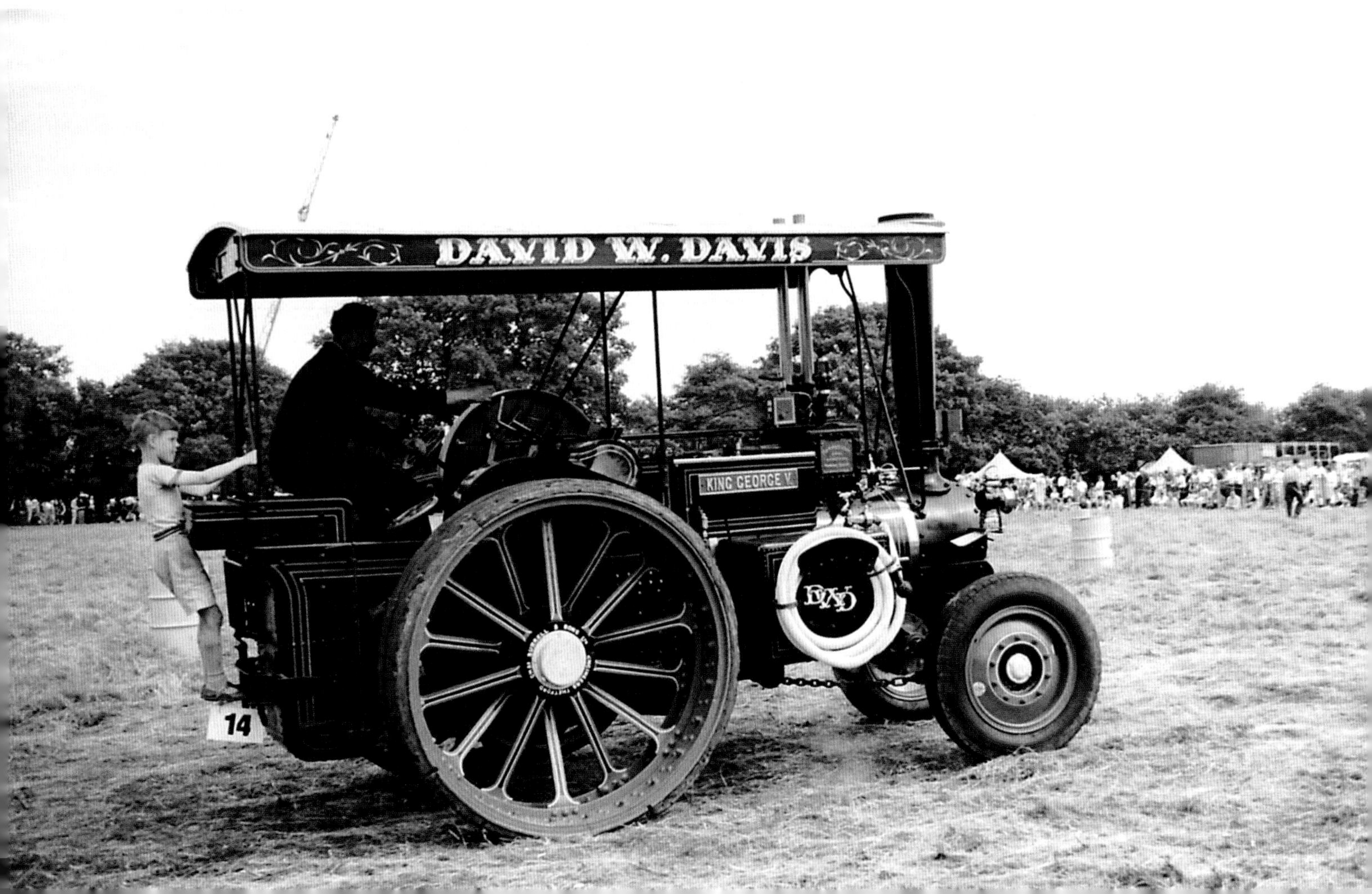
DAVID W. DAVIS
KING GEORGE V.
14

Above: A rally at Ingatestone, Essex on 17 September 1966. Visitors admire a typical Burrell agricultural engine. This is No. 3882, AH 7306 *Prince Charles*, dating from 1921. This has a single cylinder and a rating of 8nhp. This remains based in Essex but is now named *The Lurcher*. Reg Batten

Opposite above: Also present was Burrell No. 3636, CF 3476 *George*, a 1915 built double crank compound engine. This engine in 2020 was listed as resident in the Aberdeen area and carrying the names *Lord George* and *Finnola*. Reg Batten

Opposite below: The first traction engine I photographed was not actually at a rally but at an Open Day held at Cricklewood Motive Power Depot on 12 July 1969. This is Clayton & Shuttleworth No. 48154, NM 1161 a 7nhp engine dating from 1918. This was entered at the first Herts Steam Preservation Society rally in 1964 and has been a regular entrant since then.

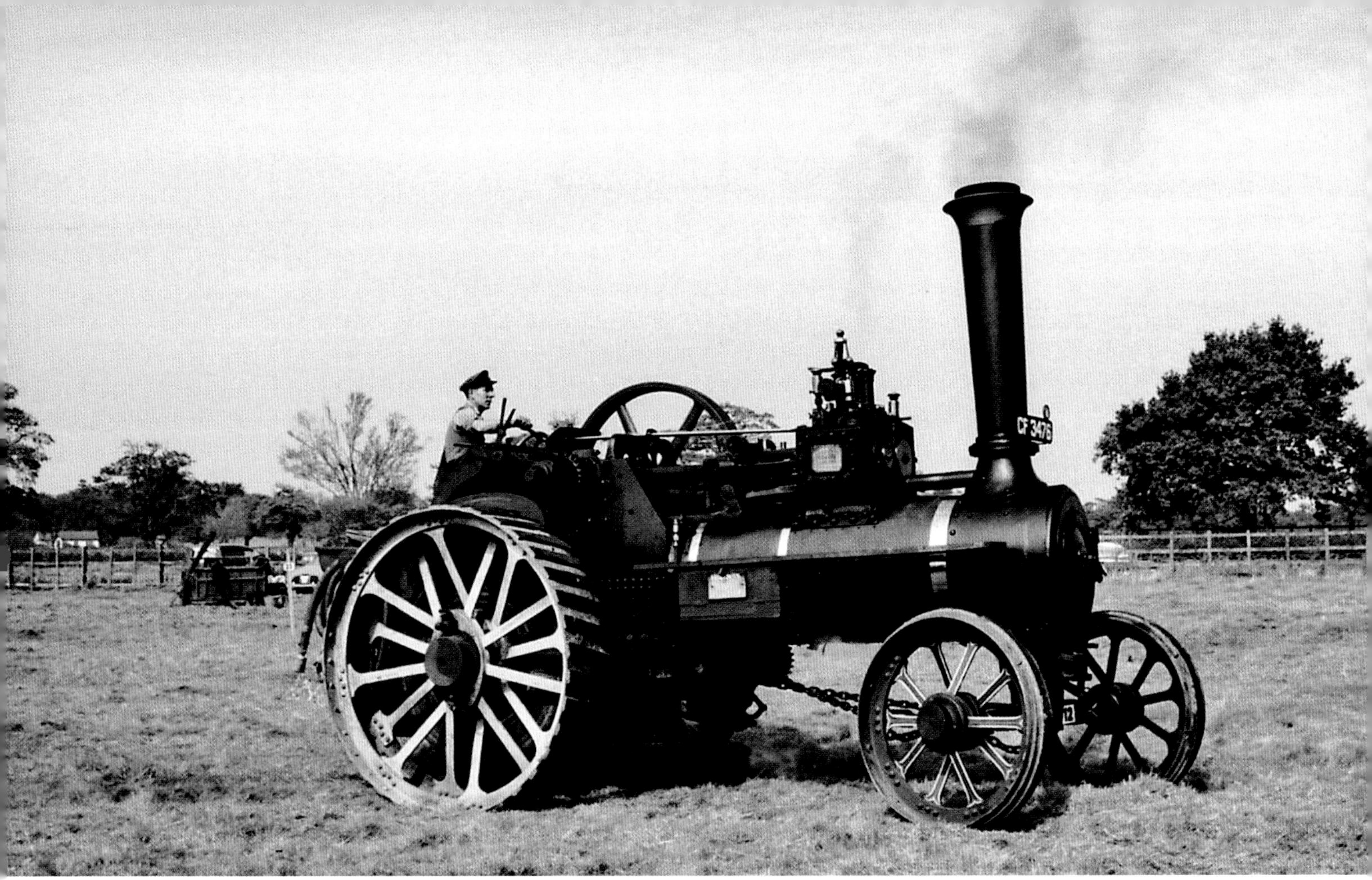
CF 3476

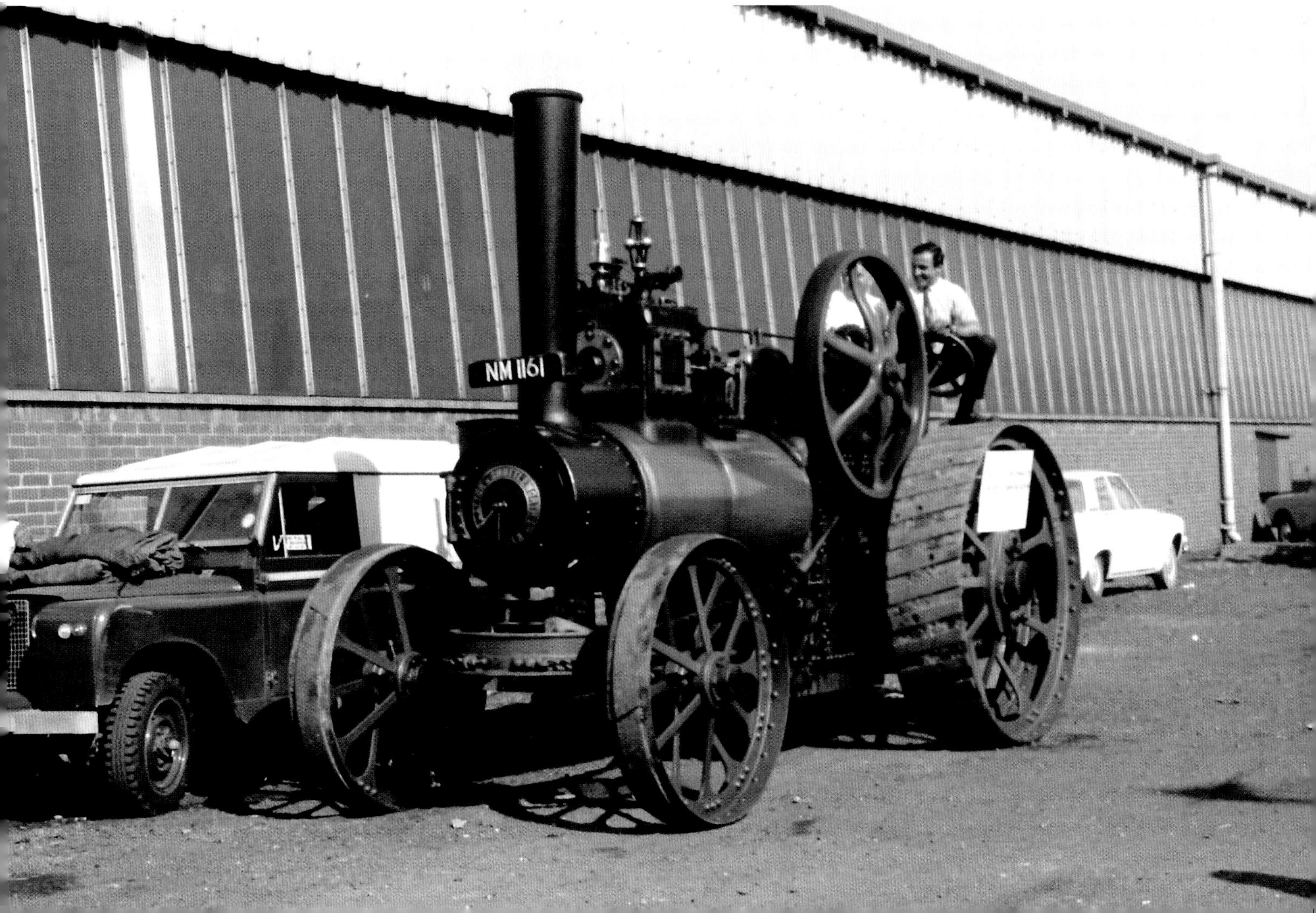
NM 1161

'ONE-OFF' RALLIES

GRAND REUNION OF BURRELL'S STEAM ENGINES 1972

In recent years there have been a number of rallies with a special theme e.g., featuring a particular marque. Such themed rallies were far less common in the early days when, after all, there were fewer engines restored. But one rally from the early 1970s was unusual in being dedicated to the engines of one manufacturer. This rally took place at West Park, Chelmsford, in Essex on the weekend of 9-10 September 1972. The rally was labelled as 'The Grand Reunion of Burrell's Steam Engines', thus it was to predate the Great Dorset's Burrell display in 2000 by 28 years.

The programme's introduction was written by the late Steve Neville, who in 1980 became the owner of Burrell road locomotive KE 3865 *Duke of Kent*. According to this, the aim of the reunion rally was to commemorate the memory of the Burrell company, the last engine of which was completed fifty years earlier in 1932. Burrell's works had actually closed in 1928, and the last few engines were completed by Garrett's of Leiston, who were a fellow member of the Agricultural and General Engineers (AGE) combine set up in the 1920s. Burrell's St. Nicholas works at Thetford, together with all its effects, was sold by auction in the autumn of 1932.

From this programme, it appears that a creditable thirty Burrell engines were booked to appear at the rally. There were twelve agricultural engines, ranging from the oldest surviving, No. 748 *Century* of 1877 to No. 4088 *Rosemarie* of 1930 – the last engine built at the Thetford works. There were also two steamrollers, one road locomotive, nine showman's road locomotives, four tractors, and one showman's tractor. Thus, there was quite a good cross-section of the company's products. The two surviving Burrell ploughing engines, Nos. 776-7 of 1879 vintage, which are now kept at the Museum of East Anglian Life at Stowmarket, had not been restored at this time.

As well as the Burrells, there were also seven engines of other makes on display, making interesting comparison with the Thetford products. One of these was Fowell agricultural engine No. 108, EW 2981. As no Burrell wagons were known to have survived at the time, a Sentinel tractor was invited instead, although a Foden or similar overtype wagon might perhaps have been more appropriate.

I particularly remember this rally for two reasons – it was the first rally I visited with a camera and also the first occasion on which I used both colour and black & white films (having only taken black & white until then).

An early Burrell traction engine, No. 1426, AJ 6705 *Old Faithful*, built in 1889. This had been brought to the rally by its owners from Yorkshire. The Albion lorry behind would make a fine entry at any modern rally too! Also in the picture can be seen two showman's road locomotives, WT 8606 *Ex-Mayor*, in its distinctive dark blue livery, stands ahead of CO 4485 *Dragon/Pride of the Fens*.

An engine which has been a regular rally attender, and looks the same today, is tractor No. 4072 *Tinkerbell*, registration number PH 2900, built in 1927. This is an example of the 5 ton-tractor and was in fact the penultimate tractor built by Burrells. The type was known as the 'Gold Medal' tractor following a trial for commercial vehicles organised by the Royal Automobile Club in 1907 at which the Burrell entry was the outright winner in the 5-ton steam tractor class winning a gold medal. The centenary of this was celebrated at the 2007 Woodcote Rally at which eighteen of the thirty-six known to have survived were booked to appear, including one now based in Holland.

Above: Burrell showman's engine No. 4000, WT 8606 *Ex-Mayor* is well known today as a member of the Saunders collection. In 1972, it was owned by William McAlpine of Henley-on-Thames (later Sir William). Here it is seen loaded ready for the journey home. This was new to Mr G. Tuby of Doncaster. He named his engines after his various civic offices hence *Councillor, Alderman, Mayor* and finally *Ex-Mayor*.

Below: Only seven engines survive from the small company of C.J. Fowell & Co. Ltd, St. Ives in Huntingdonshire. The most recent of these is No. 108 an 8nhp engine from 1922. This was owned by Gordon Wells from Dagenham, Essex at the time.

CLAPHAM COMMON, LONDON 1973
This was not actually advertised as a traction engine rally. The Bus of Yesteryear Rally organised by the London Bus Preservation Group first took place at Stratford-upon-Avon in May 1970. The rallies of May 1971 and 1972 were held in London on the South Bank and Somers Town goods yard respectively. In 1973 it moved to a weekend in September on Clapham Common. This was to form part of the London Autumn Trades Fair sponsored by the South London Press. As well as buses the event featured preserved cars, commercials, motorcycles and traction engines.

IT'S ALL HAPPENING AT THE FOURTH BUS OF YESTERYEAR RALLY

included as part of the London Autumn Trades Fair

★ Largest gathering of vintage buses at any one place
★ Steam engines and wagons
★ Veteran and vintage cars, commercials, motor cycles, etc.
★ Generous fuel allowances to all competitors, plus cash prizes of up to £50 in each class

CLAPHAM COMMON
SATURDAY & SUNDAY
SEPTEMBER 15 & 16

Entries and trade enquiries should be addressed to:
The London Bus Preservation Group,
Flat 4, 158 Windsor Road, Slough, Bucks.

1914. 6 TON. YORKSHIRE
THE CHART CARRYING Co.
BRASTED. KENT.
U-2749
THE CHART CARRYING Co.
BRASTED. KENT.
Telephone
BRASTED.
256.

166
RB3525
CHARRINGTONS TRANSPORT
Foden

Opposite above: The 1973 Show held on Clapham Common included Yorkshire wagon U 2749 of 1914. The Yorkshire wagons featured a transversely mounted front boiler, perhaps anticipating the design of the Austin Mini and similar cars by many years! 15 September 1973.

Opposite below: Another wagon at Clapham Common was 1931 Foden No. 13848, RB 3525. New to Derbyshire Council with a tipper body and here in Charringtons Transport livery, I would subsequently photograph this with two quite different bodies. (see p. 54 and p. 91)

CLAPHAM COMMON, LONDON 1977

I visited another rally held on Clapham Common on 28-29 May 1977. This was known as the Jubilee Steam Rally and was the National Traction Engine Club's tribute to Her Majesty the Queen in her Silver Jubilee year. This was the largest steam rally held in London for several years, with over fifty engines advertised, and the entries were hand-picked by a panel of NTEC members led by Jack Wharton, President of the NTEC from 1971-5.

Below: This is McLaren No. 435 *Himself* of 1892, registered in Ireland as FI 5. By 2016, this engine had returned to ownership in the Irish Republic.

Above: Aveling & Porter roller No. 11423 *Smokey* was built in 1926 as a tractor but during its working life was rebuilt as a roller with an Allen front end. This has since been rebuilt (c. 2005) as a tractor again.

Left: In 1975, two replicas were built by Belmec International Ltd of an 1865 Savage Bros chain drive engine design and one of these appeared at the 1977 show. OBW 866P was given works number 904, following on from the last surviving Savage centre engine, No.903 of 1934. Note the primitive steering arrangement with the steersman perched at the front. William McAlpine provided the drawings having acquired the Savage archive and purchased this replica.

Right: Another interesting entrant was this unregistered 1923 Leyland F5 wagon. This was found derelict in Australia, shipped home in 1968 and restored over a two-year period by apprentices at Leyland Motors. This is now housed in the British Commercial Vehicle Museum at Leyland. Founded as the Lancashire Steam Motor Company in 1896, by 1905 they had started to build petrol engine vehicles and steam wagon production ceased in 1926.

NORTHAMPTON 1996

Another rally devoted to one marque was held at Northampton on 28 April 1996. This was the home of Wm. Allchin Ltd. at the now demolished Globe Works. The company closed down in 1931, not wishing to convert to the internal combustion engine.

Below: Allchin engines line up at the rally field. Nearest is No. 1311, HR 5059 *Lena* of 1911. Beyond this is No. 1261, AY 9879 *William* of 1903. On the far left is No. 1458, DD 2006 *Jane* of 1909 which has been fitted with a canopy in preservation.

Above left: No.1546, AP 9079 *Rebel* is a 7nhp general purpose engine dating from 1912, and resident locally in Northampton.

Above right: The most recent of the surviving engines is No. 3858, first registered in 1931. NV 100 *Knapp* was built as a roller but converted as a traction engine c. 1977. The Traction Engine Register records this as having been exported to France in 2012.

LINCOLN 2000

This was not the annual rally at the Showground, but a 'one-off' road run on 1 October 2000. I didn't actually know about this in advance. I had travelled to Lincoln on a steam train excursion from London and discovered the road run on arrival – an added bonus! The road run was for engines from Ruston, Proctor & Co. Ltd and Ruston & Hornsby Ltd. Ruston, Proctor & Co. merged with Richard Hornsby on 11 September 1918 to become Ruston & Hornsby Ltd. Production was at the Sheaf Ironworks, Lincoln where the company had its origins in 1840, becoming Ruston, Proctor & Co in 1857.

Above: CT 3949, a 1907 built Ruston, Proctor & Co. Ltd single cylinder 7nhp engine, No. 33189.

Right: EB 8164 is a 5nhp Ruston, Proctor dating from 1917, makers number 51737.

Left: Ruston & Hornsby 4nhp tractor No. 52607, HP 2201 of 1918 with a trailer. Ruston & Hornsby continued the existing works number sequence.

Below: This was followed by another 4nhp tractor, No. 52573, TF 8240 *Lucifer*, towing a portable engine by the same makers. This type of tractor was known as the 'Lincoln Imp'.

FAWLEY HILL 2014, 2016

Fawley Hill, near Henley-on-Thames was the home of the late Sir William McAlpine, who was responsible for saving *Flying Scotsman* and bringing her back from America after original owner Alan Pegler went bankrupt. He also collected a vast amount of railway artefacts and installed a railway line at the estate. In addition, he also owned several traction engines over the years. A number of events were staged here, but not on a regular basis. I visited on two occasions, in 2014 and again in 2016 when a variety of traction engines were present. The 2016 Vintage Festival was to celebrate Sir William's 80th birthday.

Below: In 2014, this is No. 537, built in 1891 by Savage Bros. Ltd of Kings Lynn. They primarily built fairground equipment including centre engines for carousel roundabout rides and organ engines. This is a lighting engine i.e. a portable driving a dynamo to generate electricity to drive and illuminate a ride. This is one of two such Savage engines that survive and had once been owned by Sir William McAlpine. It would have originally been pulled to sites by horses.

Left: Among the wagons attending in 2014 was Sentinel S4 No. 8980 of 1934. This had brought over from its home on the Isle of Man for the occasion.

Below: Fowler 'Super Lion' showman's road locomotive No. 20223 *Supreme* was for many years owned by Jack Wharton. It was initially passed to the National Motor Museum at Beaulieu after his death in 1994. They offered it back to Jack's son John in 2002. It was presented at Fawley Hill in the road locomotive form in which it worked in later life.

Right: Another former showman's road locomotive that was being rallied in road locomotive guise was Burrell No. 3443 *Lord Nelson*. It was modified as such in 2000.

Below: These two engines were brought down from Banks, near Southport. MA 8472 *Little Dorothy* is Burrell No.3862. Built as a tractor, it had been converted to a roller during its working life and then converted back in preservation. The other engine is Tasker No. 1296, a class A1 3-ton tractor of 1902. This was bought by the RSPCA to assist horses in making the steep climb up the hills around Crystal Palace. It carries the appropriate name *The Horse's Friend*.

Above: Making a first appearance after restoration at the 2014 event was Aveling & Porter 1911 GND type 4nhp tractor No. 7271 *Marjorie*. New to C.F. Nash & Co. of Tunbridge Wells, by 1918 it had moved to the Peterborough Gas Company for whom it worked until 1955 before being sold into preservation.

Left: Aptly named *Colossus*, this McLaren tandem cylinder 10nhp road locomotive, No. 897 was delivered new to Argentina as a straw burner, used for pampas threshing. It was converted to diesel drive from 1948-64. The engine was repatriated and rebuilt using the original cylinders and other parts but requiring a new boiler, firebox and motion. Owners are John and Jo Atkinson of Launceston.

Right: In 2016 a most unusual roller making its first rally appearance was this vertical boiler tandem roller built by the Iroquois Iron Works, Buffalo, New York State in 1920. This was imported in 2006 and restored by Mark & Luke Wingfield.

Below: Making one of its first public appearances was *Onward*. The original Fowler 10nhp Super Lion showman's road locomotive was built in 1933 and cut up in 1953 after breaking its back axle. This replica was built by Dave Eves from 2002-16 from the original works drawings and photographs, with everything made from scratch including the nuts and bolts.

Steam fire engines are included in The Traction Engine Register and sometimes feature at rallies. At Fawley Hill in 2016 was this c.1880 horse-drawn Shand-Mason engine. Bought new by the Metropolitan Water Board it was entered by the London Museum of Water & Steam at Kew Bridge.

ANNUAL RALLIES THAT HAVE CEASED

BIRMINGHAM MUSEUM OF SCIENCE & INDUSTRY C.1969–97

The now closed Birmingham Museum of Science & Industry was located in Newhall Street, an inner-city side street a few hundred yards from the city centre and close to Snow Hill station. Each year from at least 1969 onwards, one Sunday in May, there used to be a gathering of traction engines at the museum. Around twenty or so engines would line up in Newhall Street facing the museum building. In the afternoon they would go for a short run 'round the block', a circular trip around the neighbouring back streets. This was no landscape of fields and cottages but a typical inner city industrial scenario of small workshops, warehouses and railway arches. Not perhaps an appropriate backdrop for agricultural engines, but the road locomotives and steam rollers seemed quite at home against a background that had probably not changed substantially in the last fifty years.

I went to Birmingham on a number of occasions in the 1980s for this event, which conveniently usually coincided with a bus rally at West Bromwich, so I could kill two birds with one stone as they say. The Birmingham Museum of Science & Industry was home to some engines of its own, including 1892 Aveling & Porter roller No. 2992, and 1894 Ruston Proctor portable No. 18188. On certain days, including the rally Sunday, these would be steamed while remaining inside the museum building, the smoke being extracted by chimney extensions passing through the roof. I believe this was the only museum where this took place – some other museums run their steam exhibits on compressed air. I wonder what the Health & Safety people would say about such practices these days! The museum also housed a wide range of other transport exhibits, including aircraft, motor cars, motorbikes, bicycles, and the ex-LMS Stanier 'Coronation' class Pacific locomotive *City of Birmingham*. There was also an extensive collection of machine tools, representing the strong manufacturing heritage of the West Midlands.

I last visited the gathering in 1992 – by this time photography on the road run was becoming more difficult owing to the number of parked cars and 'For Sale' notices. A few years later, in 1997, the museum was closed down, to facilitate the development of a new museum, and partly because of budget cuts.

A new interactive museum complex, known as 'Thinktank' opened on 29 September 2001 at Millennium Point, Curzon Street, after five years of construction. Some of the exhibits are displayed at this new location including Foden wagon No. 848 of 1904. Unfortunately, the steam line-up and road run have not been continued at the new location, bringing an end to what was probably the only regular opportunity to photograph traction engines on a run in an industrial street setting.

Meanwhile, the area around Newhall Street has been smartened up and rebranded as the 'Jewellery Quarter' – after one of the light industries that used to predominate there.

Above: Amongst the engines at the gathering in 1984 and a regular at Midlands events was this early McLaren No. 127, HO 5618 *Little Wonder* dating from 1882. This had also attended here as early as 1969. The engine was originally purchased for preservation by NTEC Chairman John Crawley and later owned by the late John Mayes.

Opposite above: Also in the line-up opposite the museum was Aveling & Porter roller No. 14062 *Amy* of 1930. This is an engine I may well have seen in its working days as it belonged to the County Borough of East Ham where I grew up and I can recall sometimes seeing a steam roller trundling along while on my way to school.

Opposite below: An interesting entrant in 1984 was POA 987, a home-made steam delivery van. This was created in 1973 using a 1905 van body and a 1901 Locomobile steam car engine coupled with a vertical boiler.

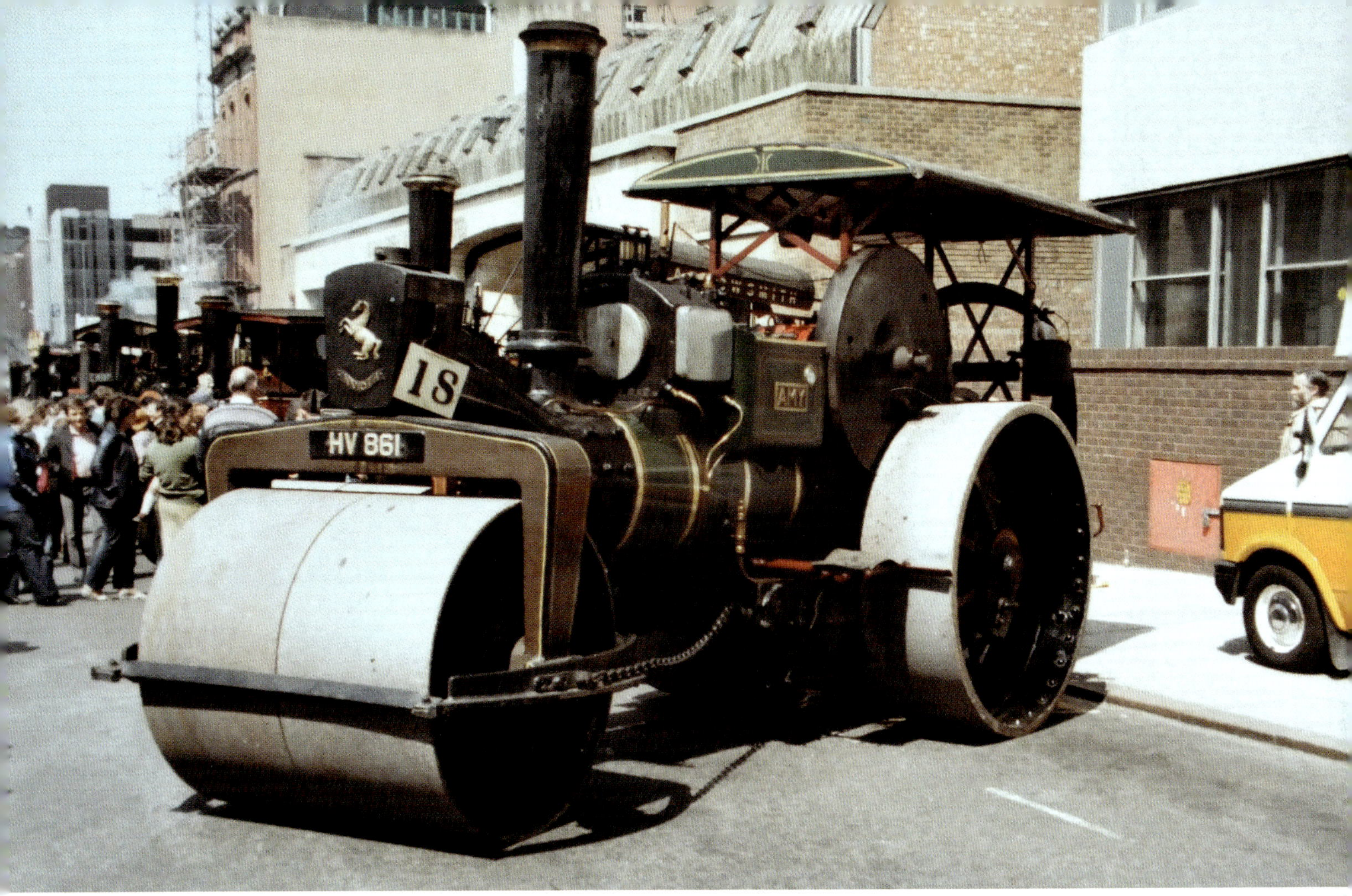
18
HV 861
AMY

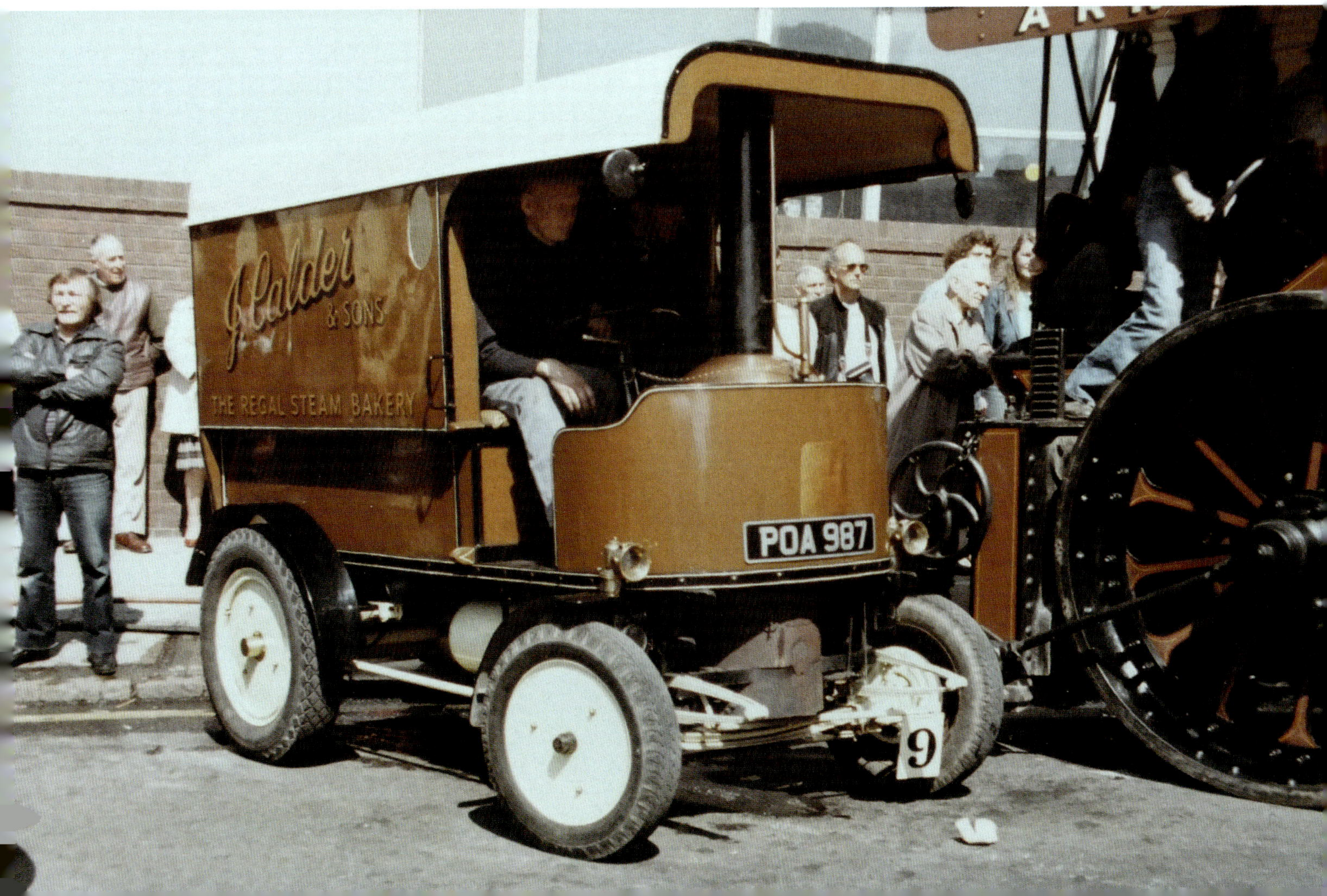
J. Calder
& SONS
THE REGAL STEAM BAKERY
POA 987
9

Above: Looking quite at home against the background of railway arches, Sentinel KF 6482 participates in the run in May 1984. Dating from 1931 this worked in the Liverpool area until passing into preservation in the early 1950s. In 1977 it was painted in the colours of Morris, Shrewsbury who once used similar wagons. For an earlier livery see p. 89.

Opposite above: Aveling & Porter steam roller No. 11520, MK 5128 *Crusader* on the run in 1985. Like many of the engines present it was based in the West Midlands, in this case at Wolverhampton.

Opposite below: Foster showman's road locomotive No. 14496, DH 4593 *The Leader* is being prepared to generate power for the organ in Newhall Street, 11 May 1986. This was built in 1921 for Pat Collin's Amusements and was in continuous service until 1958.

WOLVERHAMPTON
COWARD
ROPES
CORDS
AND
PAPERS
TWINES
HUNT LTD.
MK 5128

PAT COLLINS' AMUSEMENTS ON TOUR
DH4593

Above: Burrell 1924 built road locomotive No. 3996, E 9597 leads Fowler roller TK 6488 through a street scene of warehouses and light industrial premises. May 1986.

Left: One of the exhibits housed at the museum was this 1894 Ruston Proctor portable engine No. 18188.

Right: In steam inside the museum is resident Aveling & Porter steam roller No. 2992, AB 9331. Note the smoke extractor extension from the engine's chimney.

Below: 1931 built Fowler steam roller No. 19049, TK 6488 *Bacchus* trundles past the railway arches on 13 May 1990 with a trailer and living van in tow. A scene that might have come from 1950s, only spoilt by the glimpse of a more recent car and the modern style 'To Let' notice on the wall.

KNOWL HILL STEAM RALLY 1970–2004

The Knowl Hill Steam Rally, between Maidenhead and Reading, was founded by the late John Keeley to raise funds for building a new village hall. This was successful, the first rally attracting nearly thirty engines, and continued to grow. Over the years, the Steam Rally Trust also funded several other community enterprises. Mr Keeley died in 1999 but his widow continued to make the land available for the rally to take place. However, after the 2004 event, Trust Chairman Clifford Joseph announced the end citing 'considerable risk arising from unforeseeable weather conditions for a rally that was now costing nearly £100,000 to stage and that, even in a good year, the surplus generated did not make up for the considerable risk to charitable funds and the immense amount of voluntary work put in by supporters.' He told *Old Glory* magazine ' The increasing demands of Health & Safety requirements means the cost of staging the rally can only increase and there is now strong competition from other steam rallies and like events.'

Above: A regular feature of Knowl Hill was that Jack Wharton's Fowler showman's road locomotive No. 20223 *Supreme* would lead the ring parade. It was entered at the first rally and here it is doing just that at the 1980 event. This was the last showman's engine built by John Fowler & Co. Ltd. and was supplied to the order of Mrs Deakin, Brynmawr with chrome plated rather than brass fittings.

Opposite above: A locally owned Sentinel Super wagon was YC 7914, built in 1929 and seen in 1979. It was originally built for use by Henry Butt & Co. in a Mendip quarry as a tipper wagon where it spent all its working life, never being fitted with pneumatic tyres.

Opposite below: The Sentinel S4 wagons were the ultimate design from this maker, with a vertical boiler and underfloor-mounted engine. No. 9293 dates from 1937. In 1980, it was carrying this livery but has since changed hands and has been repainted in the colours of the original owners, the Castle Firebrick Co. Ltd of Buckley, Flintshire.

SENTINEL 1926
46
DAVID BERESFORD
AND SON
KNOWL HILL
YC 7914

46
EARLY TRANSPORT
WAREHAM
DORSET
UJ 9497

Above: Over the years, Ruston Proctor No. 44180 of 1912 was entered in various stages of restoration. Here it is being towed around the ring in 1990. This is a straw-burning export engine that was imported from the Philippines.

Below: A Knowl Hill regular was Aveling & Porter 1914 6nhp road locomotive No. 8471 *Clyde*. This was in 1990 when the engine did not have the canopy that was later fitted. On the back of the footplate is well-known rally character, Dr Busker.

Above: Aveling & Porter ploughing engine No. 6547 of 1908 was imported from West Africa in 1993. This was shown in 'as acquired' condition at Knowl Hill in that year.

Right: Entered in 1994 was this Robey undertype semi-portable engine, No. 49116, which was built as late as 1940.

Above: Quite a beast is NK 1755 *Sally*, one of a pair of Fowler ploughing engines dating from 1917 which were converted to diesel drive with a 6-cylinder engine mounted on top of the boiler. This and NK 1754 *Salty* were still being used commercially for dredging at this time.

Opposite above: John Mr Keeley owned AB 9987 *My Delight*, a rare Fowler DD type 8nhp ploughing engine. Ploughing did not take place at Knowl Hill, so instead the engine is seen driving a band saw with the belt drive. 1996.

Opposite below: In 1996, Fowler road locomotive XC 9653 *The Lion* was painted in wartime livery and carrying the name *Lafayette*. This had been built in 1914 for the War Department to haul the big artillery guns during the First World War and was sold on in 1918 for road haulage work.

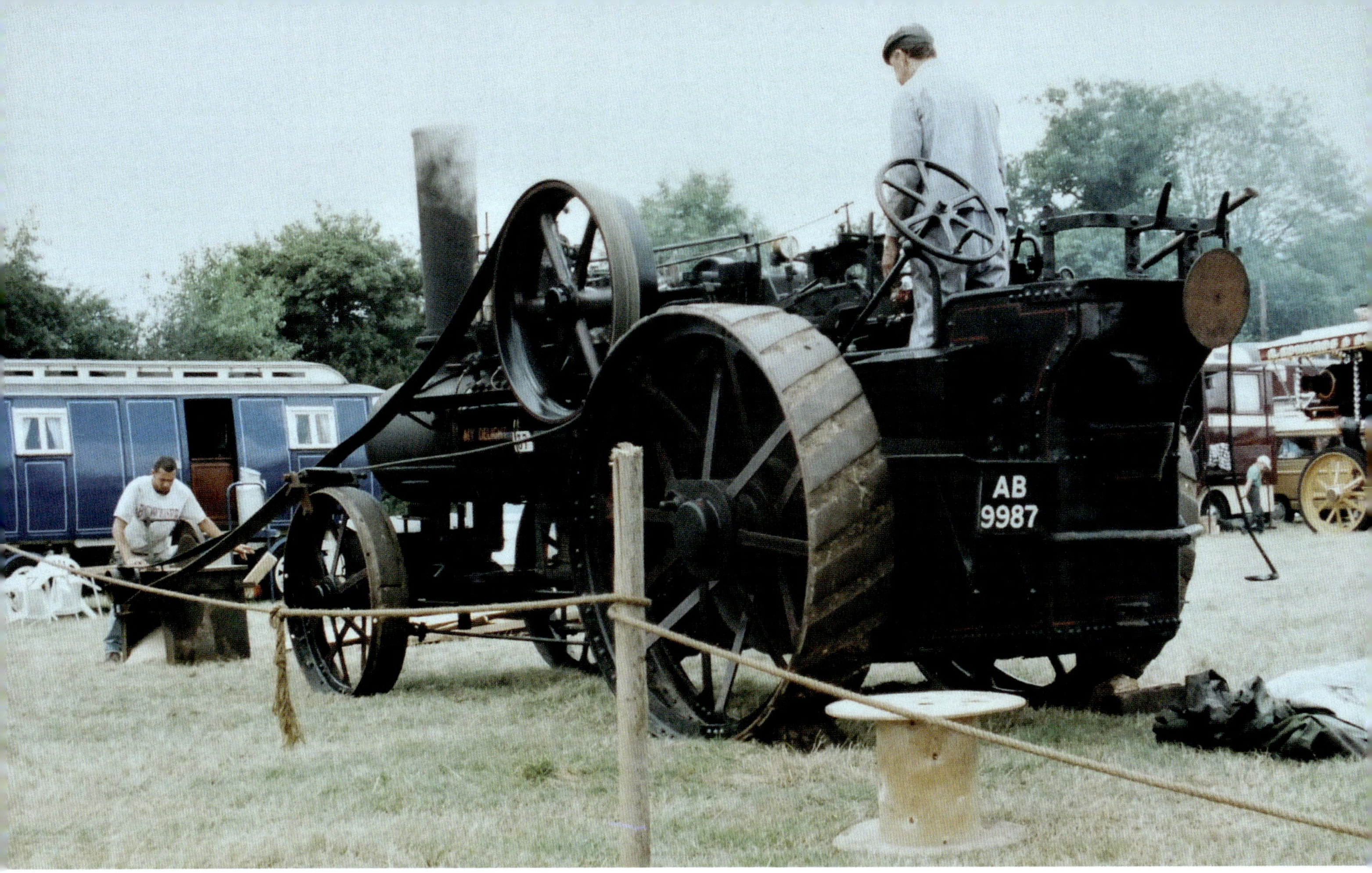
AB
9987

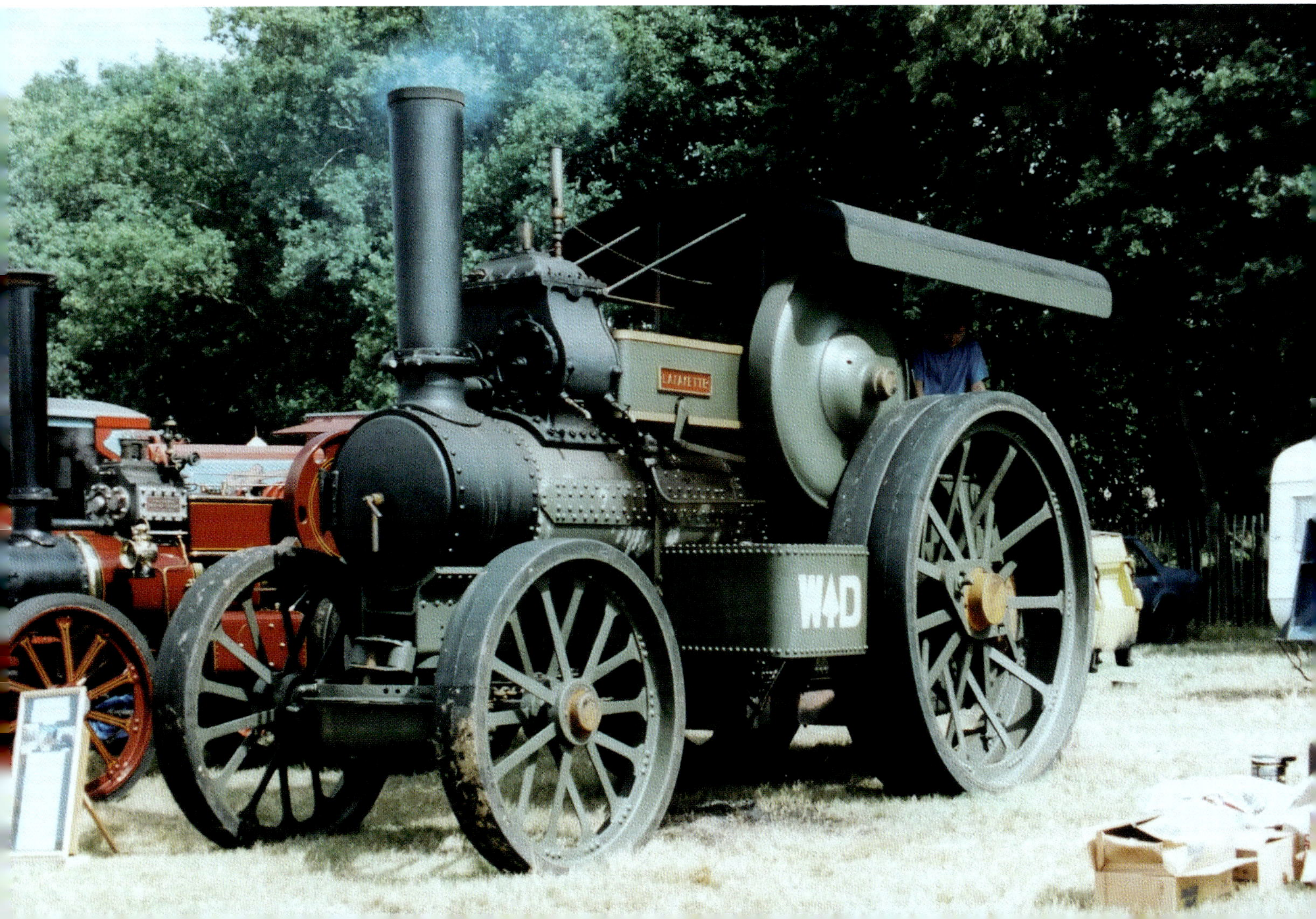

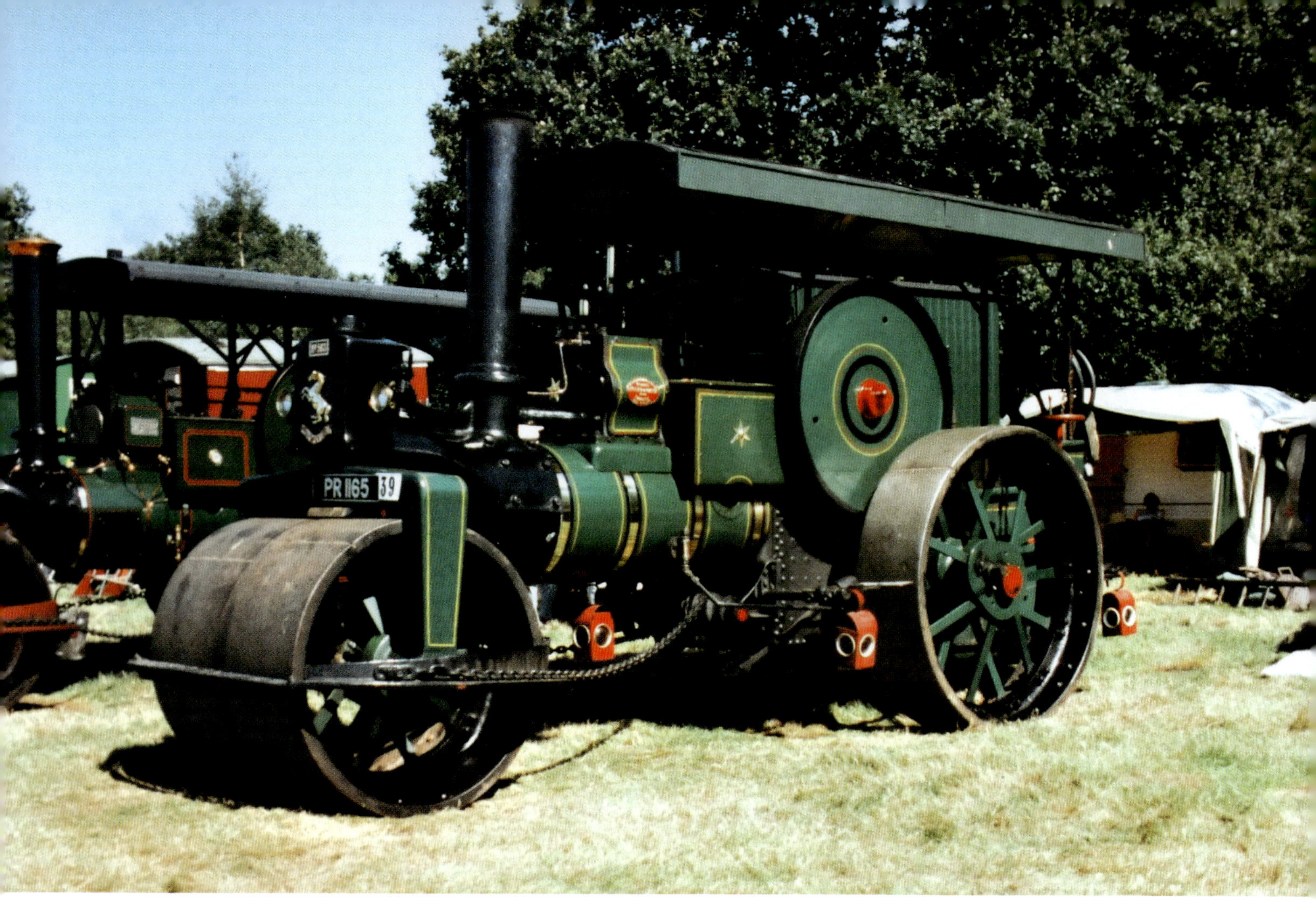
PR 1165
39

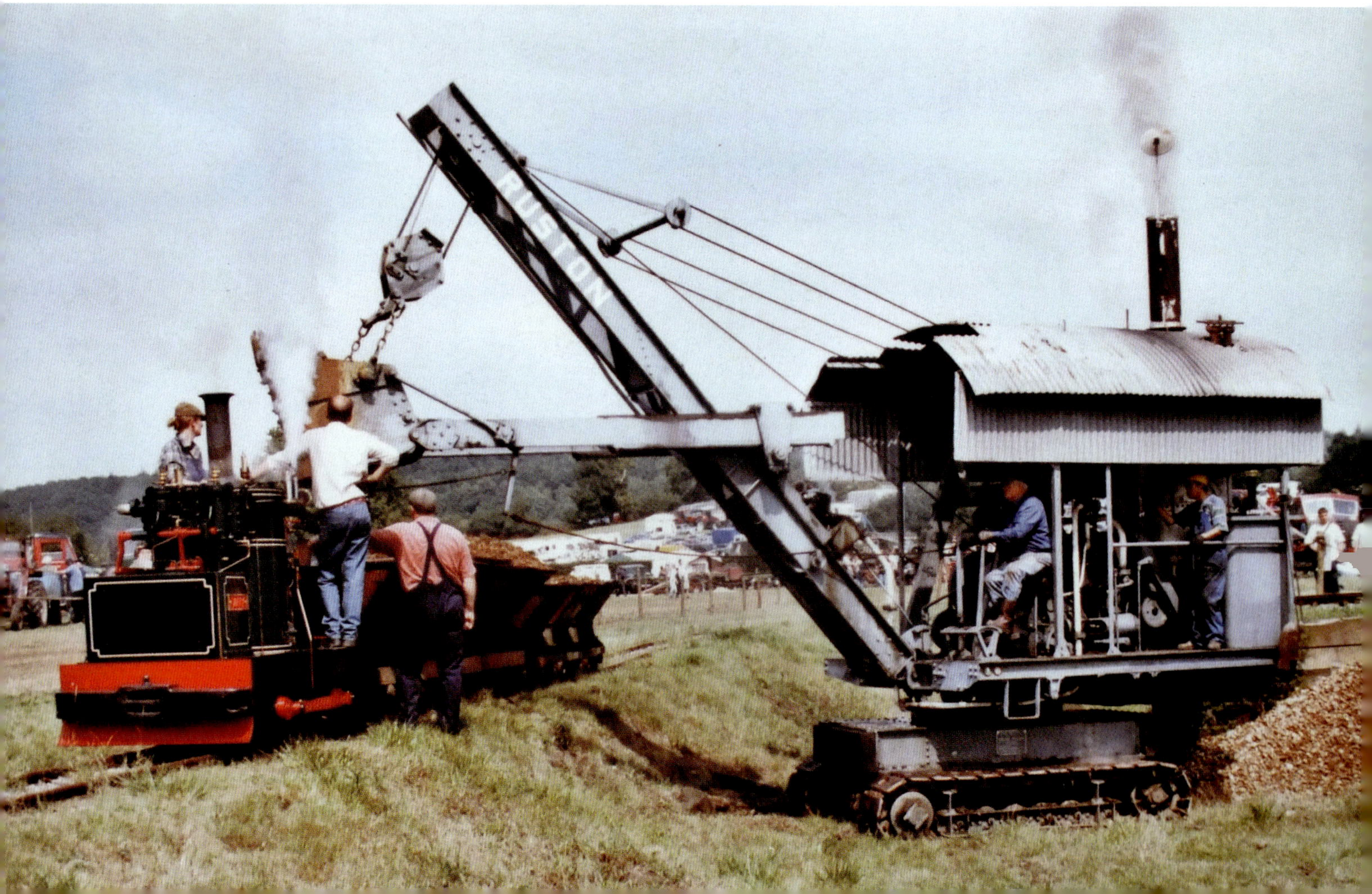
RUSTON

Opposite above: Aveling & Porter roller No. 10671, PR 1165 *Churchill* was an entrant in 1998. Built in 1923, it was exhibited at the Newcastle Show in June 1923. It was then purchased by the well-known contractors Eddisons of Dorchester. On withdrawal in 1957, it was presented to Westminster City Council to become a static feature in a children's playground on the Churchill Estate, Pimlico (hence the name). It was rescued for preservation in 1981 and restored to working order. It passed to the present owner in 1993.

Opposite below: A feature in the later years of Knowl Hill was the quarry railway. A steam crane would load up hopper wagons on a 2-foot gauge railway, after which a different visiting steam loco each year would pull the wagons to the other end of the line where the spoil would be tipped out. In 2004, this vertical boiler 0-4-0 *Taffy* was in use. Although I did not know it at the time, this would be the last Knowl Hill Rally.

LAMBETH COUNTRY SHOW C.1973–1990S

An inner-London suburban park may seem an unlikely location for a country show, but the Lambeth Country Show held in Brockwell Park is that. Organised by Lambeth Council this began in c.1973 and takes place usually on the third weekend in July. Until the 1990s this included traction engines and vintage vehicles but now is more of a community cultural and music festival that reflects Lambeth's ethnically diverse population. I first visited in 1977. In 1978, a London Bus Rally and road run was added to this, sponsored by Grey-Green Travel. This would be repeated in 1979, but the bus rally later moved elsewhere. I last visited in 1990 by which time the number of engines had declined. The Brixton riots of 1981 and 1985 may have been instrumental in the change of direction of this event to represent the local culture.

Below: On 23 July 1977 at Brockwell Park we see 1934 Sentinel S8 wagon No. 9105, UJ 3652. This was built as a demonstrator and later worked for Wynns of Newport.

Above: Also present was Foden 1931 wagon No. 13848. I have photographed RB 3525 with three different bodies – at this time it was carrying this tar tanker body. In 1973 it had carried a tipper body. (see p. 26)

Opposite above: Aveling & Porter roller FX 7014 *Moby Dick* proceeds around the ring. That tower block just visible above the trees is not the backdrop you normally find at traction engine rallies!

Opposite below: Moving forward to 1986, and this is Burrell No. 3894 *Saint Brannock*. This was supplied new in 1921 to J.B. Dugdale from Warwickshire and probably used for haulage. After sale into preservation, it was rebuilt with showman's fittings c. 1970.

FX 7014
AVELING & PORTER, Built 1913
208
4

NX 947
WILLIAMS & SONS
On Tour

EXPO STEAM, PETERBOROUGH SHOWGROUND 1972–89

Held at the East of England Showground, Alwalton, Peterborough over the August Bank Holiday, I first visited this in 1979 and continued most years until 1989. I believe it ceased after this. From 1992 to 1999 there was an annual rally held at Tallington, a few miles north of Peterborough.

Above: One of the main things I recall about this event was that whereas at most rallies the engines are parked up alongside each other, here they were parked nose to tail but well spaced-out in two long lines. This made it very easy for photography. At the 1979 Expo, this is Fowler 7nhp engine No. 18155 *Robert the Bruce*, dating from 1929 which had come from Leicestershire. Fowell engine CE 7894 of 1904 is behind.

Opposite above: From the collection of Robert Crawford, Frithville, Boston came this rare Foden 'Sun' type tractor. No. 13730 dates from 1930 and is one of only two of this type preserved in the UK and Ireland.

Opposite below: One of the many rollers present in 1979 was Fowler No. 18651 of 1930, an 8-ton DNA model.

R^t H.CRAWFORD & SON
Agricultural Engineer
FRITHVILLE
Foden
SAND. STONE
DAILY
TEL
322
GISBURN
60
WFORD
ural Engineers

VF 9312
FOWLER

Above: The 1980 show had a wagons theme, and this produced more rarities. Also from the Robert Crawford stable came this 1900 built Thornycroft wagon, makers number 39. This was originally owned and restored by John Crawley. A similar design wagon, No. 115 survives and resides at the Milestones Museum in Basingstoke. However, the company's engine No. 1, a van from 1896 also survives and this is kept at the British Commercial Vehicle Museum at Leyland.

Opposite above: Several of the wagons were from the extensive collection built up by Tom Varley. Making a first public appearance after restoration was Foden 'Colonial' No.4086 of 1913 which had been repatriated derelict from Australia the previous year. The paint was barely dry! This wagon was later re-exported to Germany where it spent 1984-2008.

Opposite below: Also present from Tom Varley's collection was FD 6603 *Pendle King*, a 1931 Sentinel DG6 wagon. Contrast its appearance with that of the same wagon when in its working days (see p. 11). Behind this is UB 8660, a 1931 Fowler wagon. He also had a pair of Yorkshire wagons with transverse mounted boilers, both dating from 1927. UA 1163 is a tractor unit from an articulated wagon and is on solid tyres, while UA 1788 is on pneumatic tyres.

CARTAGE
INTERSTATE
SYDNEY
T. VARLEY
AUSTRALIA
6 Ton Colonial
M 4848

Sentinel
ENGINEERING
King of the
PENDLE KING
Road
HAULAGE
AND
STORAGE
5
T.
VARLEY
&
SONS
MUSEUM
OF STEAM
VARLEY &
JACKSON
31
FD 6603
TOM VARLEY
GISBURN
No 3

Above: A fine pair of matching Fowler ploughing engines seen in 1980. Nos 15441-2, *Tiger* and *Lion* date from 1920. Note the different chimney styles, evidence of replacement at some stage.

Below: At the 1982 show is Aveling & Porter 1908 built tandem roller No. 6530. This is now kept at the Stockwood Discovery Centre, Luton (see also p. 108).

Above: Although the site was a showground, there was farmland alongside and ploughing could be demonstrated in a field next to the site. Fowler No. 15670 is one of four large Z7 class 22nhp engines that were brought back from Mozambique. All date from 1922.

Below: An interesting exhibit in 1983 was this traction engine built by J&F Howard of Bedford in 1872. This is the only traction engine surviving by this make although there is also a ploughing engine. (see p. 109)

RUSHMOOR ARENA 1975, 1978–98

Aldershot is very much an army town. Much of the land is taken up with barracks and other facilities. Just on the outskirts of the town, the army had a showground, where military tattoos and other events could be staged. This site, Rushmoor Arena, was the location for a popular series of steam rallies for some twenty years, but since ceased.

Rushmoor Arena was created in 1922-3 as a site to replace Cove Common which had been used since 1920 to stage the annual military Searchlight Tattoo, an event which had its origins in 1894. Gangs of local unemployed people were recruited to clear, level and turf the central area of the 68-acre site. A grandstand for 800 people was completed in time for the 1923 event. In 1922, some 22,000 people attended at Cove Common, in 1923 some 43,000 attended at Rushmoor.

The Tattoo became a major event through the twenties and thirties. In 1939, the event, now stretching to nine days, brought well over half a million spectators to Rushmoor. However, the outbreak of war brought events to an end. The Arena was used as a vehicle depot during the Second World War and then abandoned for the next twenty years, with some stands and buildings being sold in 1956.

An Aldershot Army Display was instituted in 1963. In 1972, the Arena was authorised for restoration to house this event, and then became available for other activities.

The first event at Rushmoor Arena that I attended was on 20 July 1975. It was billed as 'The Yesteryear Show'. Entries were invited for preserved buses, commercials etc to appear at this 'new event in the largest arena in the South'.

There was indeed a good selection of preserved buses and commercial vehicles present, and vehicles were mainly well positioned for photography – an important consideration as far as I am concerned! There was also a selection of traction engines present – an added bonus I wasn't expecting. Amongst those present there that I photographed were Burrell tractor AH 0563, Marshall AJ 5781 *Pride of the Road*, Sentinel wagon PD 1701, and Wallis & Steevens rollers HO 6341, NK 7051.

Looking at my pictures from this event, there did not appear to be all that many people there, but presumably the event was considered success enough to inspire the use of the site for future rallies. As a result, the 'Rushmoor Steam and Vintage Show' started in 1978. The event was organised by the Three Counties Steam Preservation Society who had previously staged some one-off events elsewhere locally.

Rushmoor was a NTET approved rally. It soon became a popular event. An average of 50-60 traction engines attended each year at the show, which was held on the last weekend of July. All the expected features of a successful steam show were present, except that ploughing was not possible. I never went to the earlier shows, but I went to all but one of the shows between 1984 and 1995.

Rushmoor Arena was set in a wooded area. Steam entrants parked up against a backdrop of trees when not in the ring. A unique feature was the terraced stand alongside the main ring. Visitors could sit on the terracing and enjoy a grandstand view of ring events. During ring displays, vehicles would enter the ring from one side, then exit from the other, after which they would face a stiff climb up a road that led round behind the back of the top of the terracing.

There was strong local support. Charlie Russell from Bordon was there with several of his collection. The Hollycombe Steam Collection at Liphook sent various engines from their stable over the years. Bicknells' Ransomes crane engine *Hooky* would regularly be at work in the wood sawing demonstration area. From somewhat further afield came the Saunders family with their showman's engines *Ex Mayor* and *Carry On* powering the organ which was accompanied by Penny Rigden's dancers.

The Hampshire County Museum Service regularly had a display area where they were able to present and run some of their extensive collection of locally built steam power and Thornycroft lorries. This offered a rare opportunity to see this collection at a time when the Milestones Museum at Basingstoke was only a pipedream. Several rare Tasker engines which were saved by the Tasker Trust and cared for by the Museum Service were rallied, including the oldest preserved, a stationary engine of 1872, and the most recent a steam roller of 1926. In 1991, the sole surviving Tasker wagon No. 1915 made its debut after a 6-year restoration.

Being on army land, there was always a good display of military vehicles, both current and in preservation. I recall that in the years after the Falklands war of 1983, several captured Argentine vehicles were on show.

However, the army was making cutbacks, and Rushmoor Arena was little used, so there was a desire to close it. The annual Aldershot Army Display had ceased in 1984, due to a lack of availability of troops owing to the pressures of commitment in Northern Ireland. I believe the terrace stand was also giving cause for concern. Anyway, presumably as a result of this, no Steam & Vintage Show was held in 1996 and 1997. However, the show returned in 1998 for what turned out to be the last time.

At the final rally held in 1998, only thirty-one traction engines attended, rather than the usual fifty plus. The terrace stand was fenced off, out of bounds. The show was all right but not of course up to usual standards. And although we didn't know for sure at the time, it was the end of an era.

Also present was Sentinel Super type wagon No. 5509 of 1924 which is fitted with a three-way tipping body. Note the lack of visitors evident in these photographs.

Above: Among the engines present in 1975 was Burrell 4nhp tractor No. 3807, AH 0563 *Nobby*. This engine has since been fitted with a crane.

Opposite above: A busy scene in the ring at the event held on 27 August 1985. In the foreground is F 5218, which the canopy tells us is Aveling & Porter No.7612 built in 1912. What it does not say is that this started out as a tractor and was later altered to have showman's fittings, although this happened prior to 1960.

Opposite below: Foden tractor No. 12370 of 1926 passes behind the ring and the terrace stand in 1987. At this time, it carried this green livery but was later repainted brown with lettering for the original owners, Arthur Kirby of Newton Abbot sawmills.

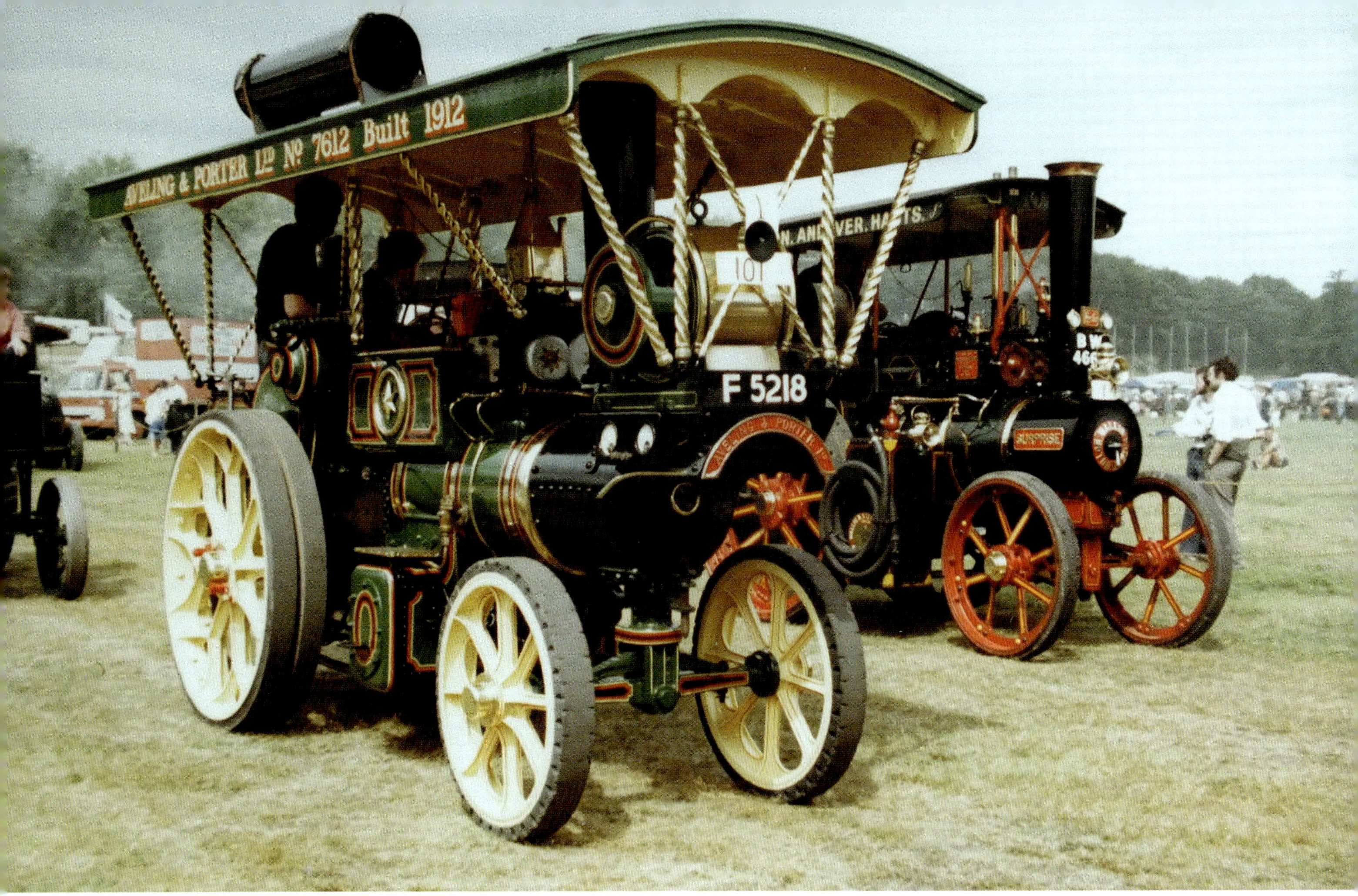
AVELING & PORTER LTD No 7612 Built 1912
F 5218

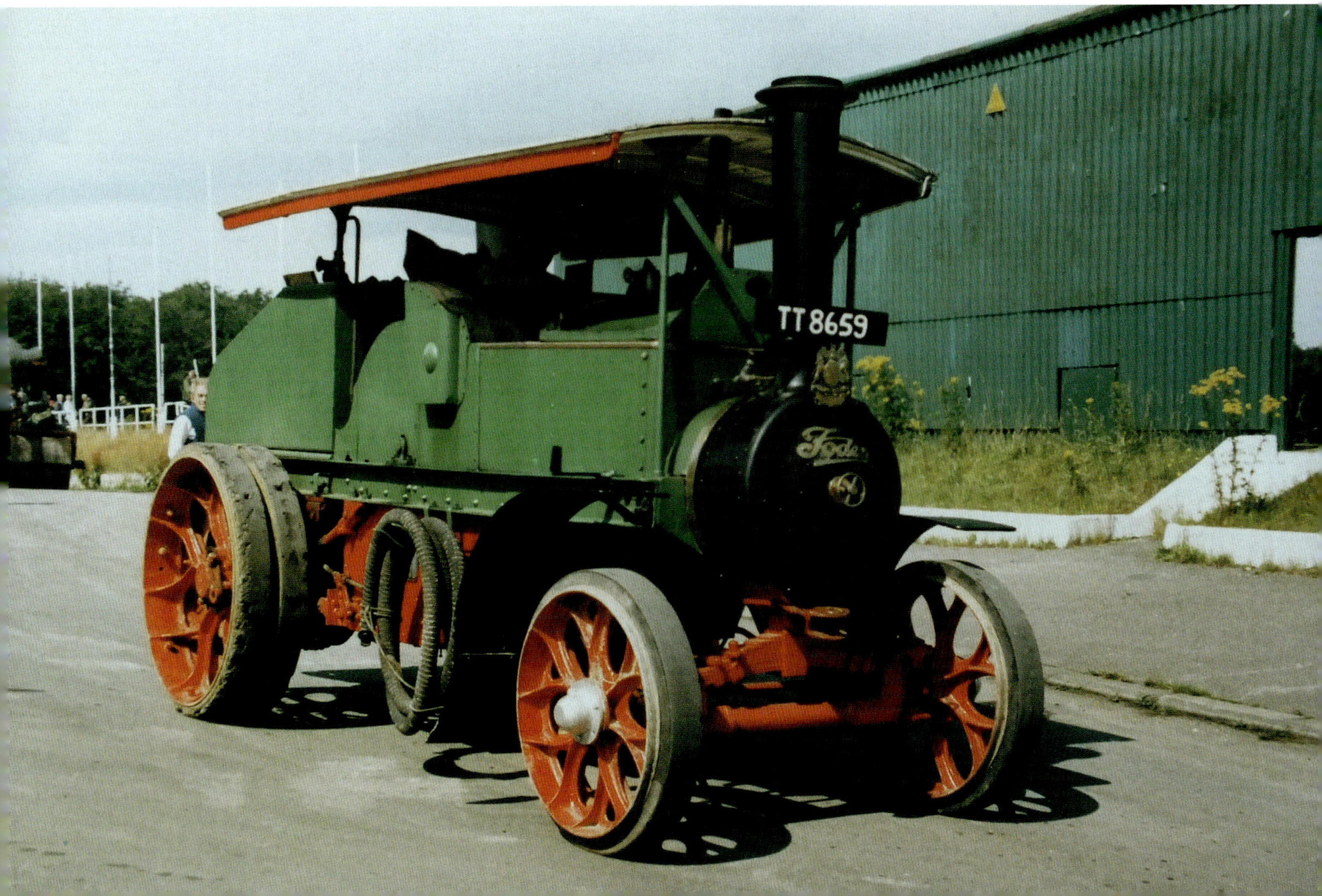
TT 8659

Above: In 1988 we see Aveling & Porter No. 7411 of 1911. This is the sole survivor of nine vertical twin-cylinder Shay-geared tandem rollers made by the company.

Opposite above: The Hampshire County Museum Service display at the 1991 show featured the sole surviving Tasker wagon No. 1915, making its debut after a 6-year restoration. Built in 1924, this worked for W.J. King, quarry owners of Somerset. It was rescued from their quarry near Minehead in 1957. Behind are Tasker stationary engine No.111 of 1872 and portable engine No. 1228 of 1898. All these were saved by the Tasker Trust and maintained by the Hampshire County Museum. Since 2000 they have been kept at the Milestones Museum in Basingstoke.

Opposite below: Ransomes, Sims & Jefferies 1891 portable engine *Island Mor* is coupled to a Dennings of Chard baler at the 1993 rally.

Hampshire
County
Museum
Service
W. J. KING.
BAGBOROUGH.
SOMERSET.
THE LITTLE GIANT

DENING'S of CHAR
SOMERSET
STRAWBALER

Above: In 1993 the Hollycombe Steam Collection entered their unique Fowler/Allen ploughing engine BW 4613. This has 'tank' steering – turn the wheel to the left to turn right, and it was a difficult job to steer this around the ring, as evidenced by the two crewmen needed to work the vertically mounted steering wheel.

Opposite above: Ransomes crane engine *Hooky* prepares for another lift in the timber sawing demonstration area at Rushmoor in 1995. Owned by Bicknells of Haslemere, this is the former Ransomes works engine. It would have been used to move heavy parts such as boilers around the Ipswich factory.

Opposite below: At the final 1998 show, driving the sawbench is 1903 Foster No.2728, V 7917. Behind it can be seen a red Thornycroft lorry and the Tasker wagon, both from the Hampshire County Museum collection.

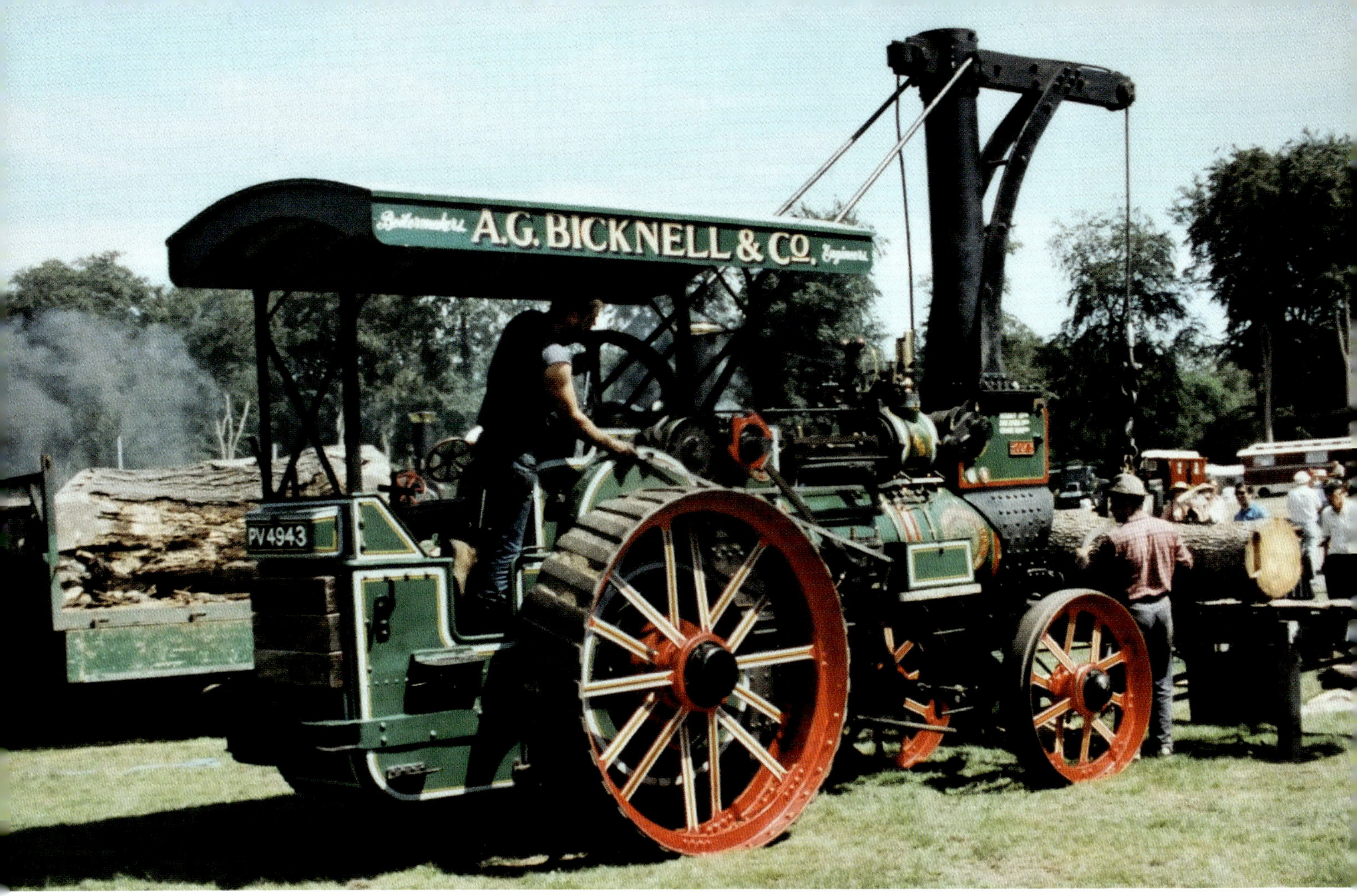
A.G. BICKNELL & Co.
PV4943

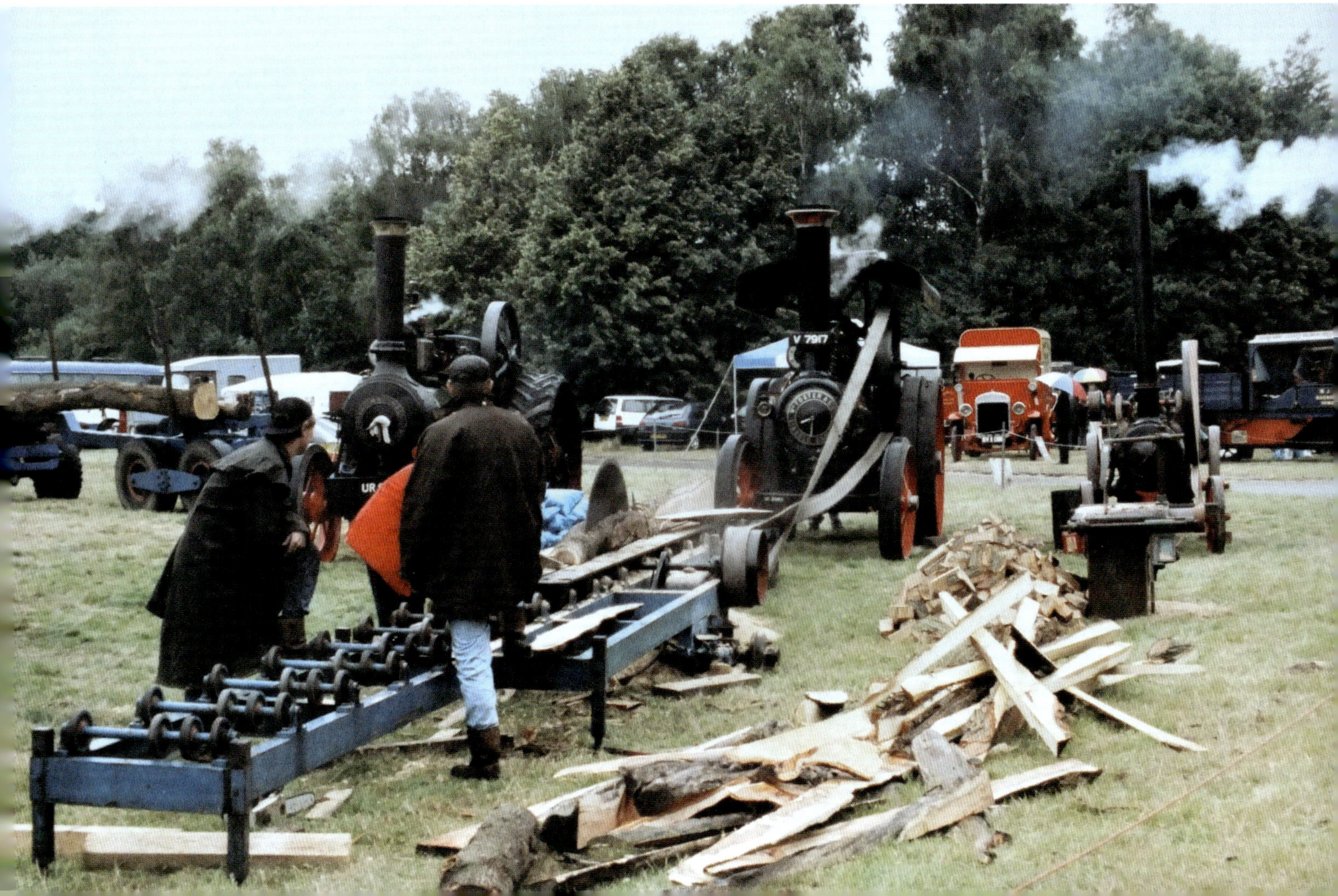

ESSEX STEAM & COUNTRY SHOW/ESSEX COUNTRY SHOW 1986–2016 (NTET AUTHORISED EVENT)

From 1986 this event was held at the Barleylands Farm Museum (now Barleylands Farm Park) near Billericay, owned by H.R. Philpot & Son. It was renamed as the Essex Country Show from 2003. The farm museum owned some resident engines either active or on display in the museum area. These included a pair of Fowler AA ploughing engines Nos 14728-9 *Giant Tiger* and *Giant Panther.* The show featured extensive demonstrations of ploughing, threshing, sawing and harvesting by steam, tractors and by horses. As this was the nearest rally to my home in east London, I visited most years from 1990 onwards. But by 2012 the numbers attending started to decline and after that I stopped going. The event was axed in 2016 due to increased operating costs.

Opposite above: One of the most significant features of the Barleylands site was its ability to stage ploughing displays. Up to four sets of ploughing engines would be on hand demonstrating with different types of equipment. Here in 1995 Fowler No. 15335 *Lady Caroline* is seen with a four-wheeled cultivator. This engine was one of those that attended the famous 'Raynham Day' rally in September 1962.

Opposite below: David Birch of Sellinge, Kent was a regular with *The Banshee*, McLaren No. 1713 of 1922. This had been reconstructed from a crane engine to a showman's road locomotive in the 1970s. After carrying the usual maroon livery for some years, in 1996 it gained this blue scheme. This has since been sold to another owner with the intention to rebuild it to a crane engine.

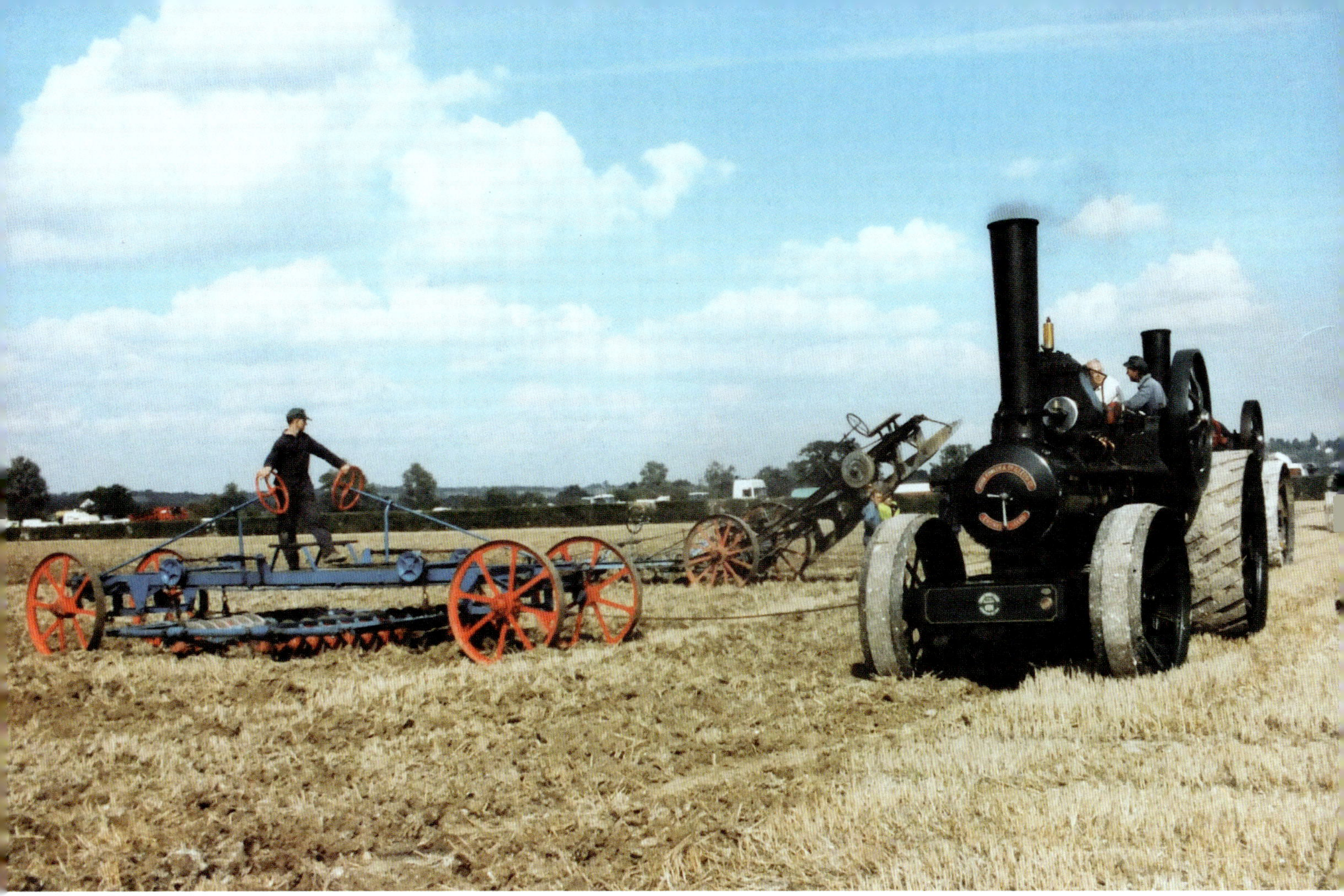

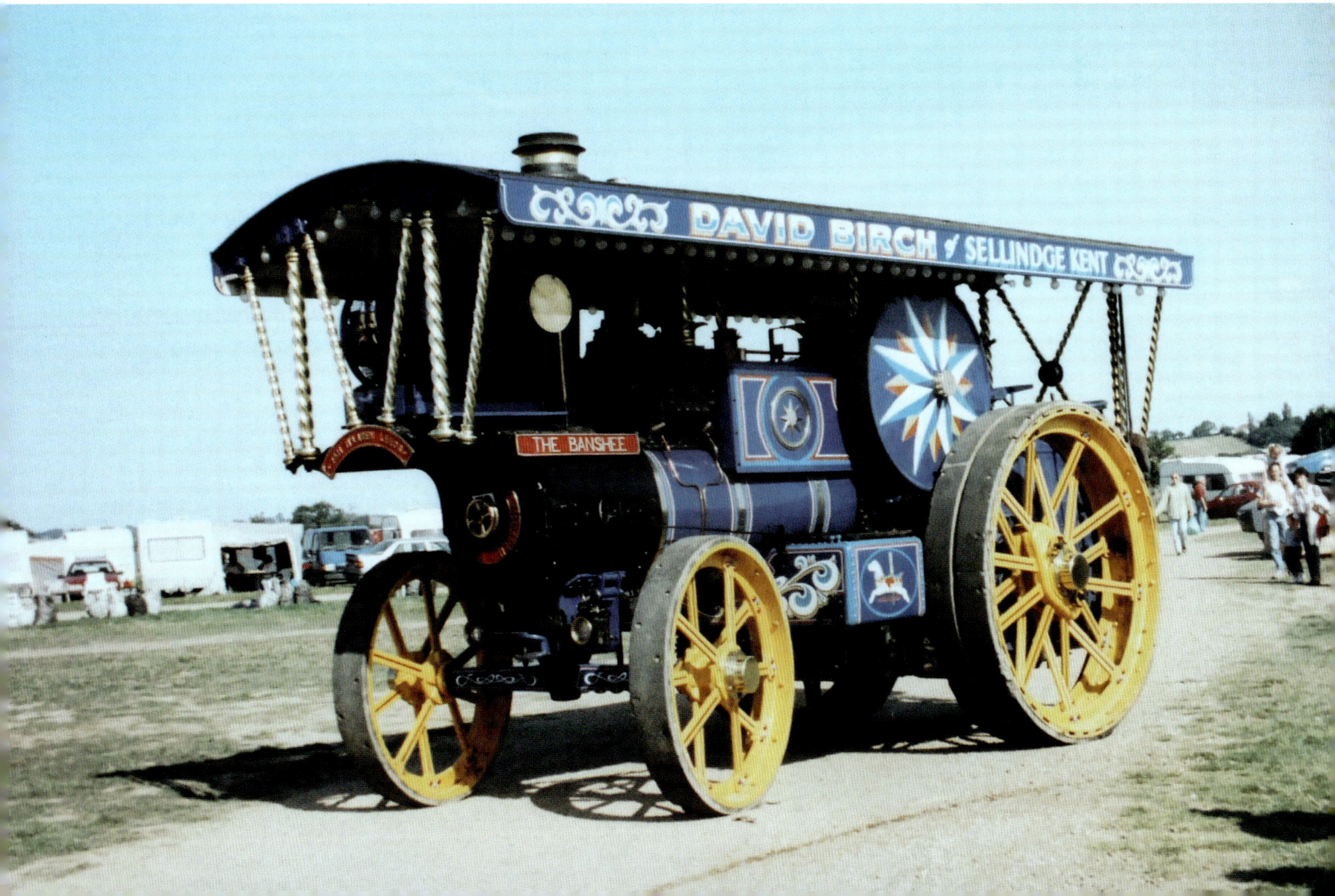
DAVID BIRCH of SELLINDGE KENT
THE BANSHEE

Above: Garrett designed the 'Suffolk Punch' tractor as an attempt to compete with internal combustion tractors at direct ploughing. This is the sole survivor of the type, seen on the ploughing field in 1997. This engine is now kept at the Long Shop Museum in Leiston, Suffolk, housed in part of the former Garrett workshops.

Opposite above: In 2000, five Burrell traction engines were shipped over from New Zealand and attended several rallies. Burrells Nos. 3229 of 1910 and 3148 of 1909 visited Barleylands and were given the opportunity to show their strength hauling a modern tractor on the low loader.

Opposite below: This Wallis & Steevens 3-ton tractor No. 2811 *Lena* was built in 1905. In 2000, it was purchased by Carolina Schreiver from Holland. She had developed an early interest in traction engines while working with Giles 'Doc' Romanes on his similar engine. Carolina was a regular visitor to Barleylands with her engine which she also rallied in France and Holland. Here she was with her engine in the sawing area at the 2001 Show.

123
THE BURRELL TRACTION ENGINE

RANSOMES STEAM TRACTOR
AD 7460

Above: Fowler AA ploughing engine No. 15563 *Wayfarer* was one of a pair that was new to Lord Rayleigh of Terling, Essex, a prominent landowner in the county. As well as ploughing, they were also used for other work on the estate farms including threshing, baling and wood sawing. They last worked as a pair in 1948, both later passing into preservation. They were acquired in a derelict state by their present owner, Alex Sharpehouse, in November 2003. Restoration of *Wayfarer* was completed in the spring of 2004. The owner comes from near Grange-Over-Sands in Cumbria and decided to bring the engine some 350 miles south to Barleylands in 2004 so that it could perform once again in the county where it spent its working life.

Opposite above: There were extensive displays of farming as might be expected at a farm museum. Horses, steam power, tractors and combine harvesters all demonstrated their working methods. Each year, Wallace & Steevens No. 7683 *Eileen the Erring* could be found in the same place, powering this 1949 Ransomes threshing machine. This was it in 2004. This was a locally based engine, coming from Tilbury. It was this engine, then owned by Giles Romanes and just called *Eileen*, that unsuccessfully challenged Arthur Napper's *Old Timer* in the 1951 race at Nettlebed.

Opposite below: There was plenty of opportunity for the miniatures to get involved. This half-scale Case engine *Kansas Glen* was driving a sawbench in 2005. This was a regular visitor to Barleylands.

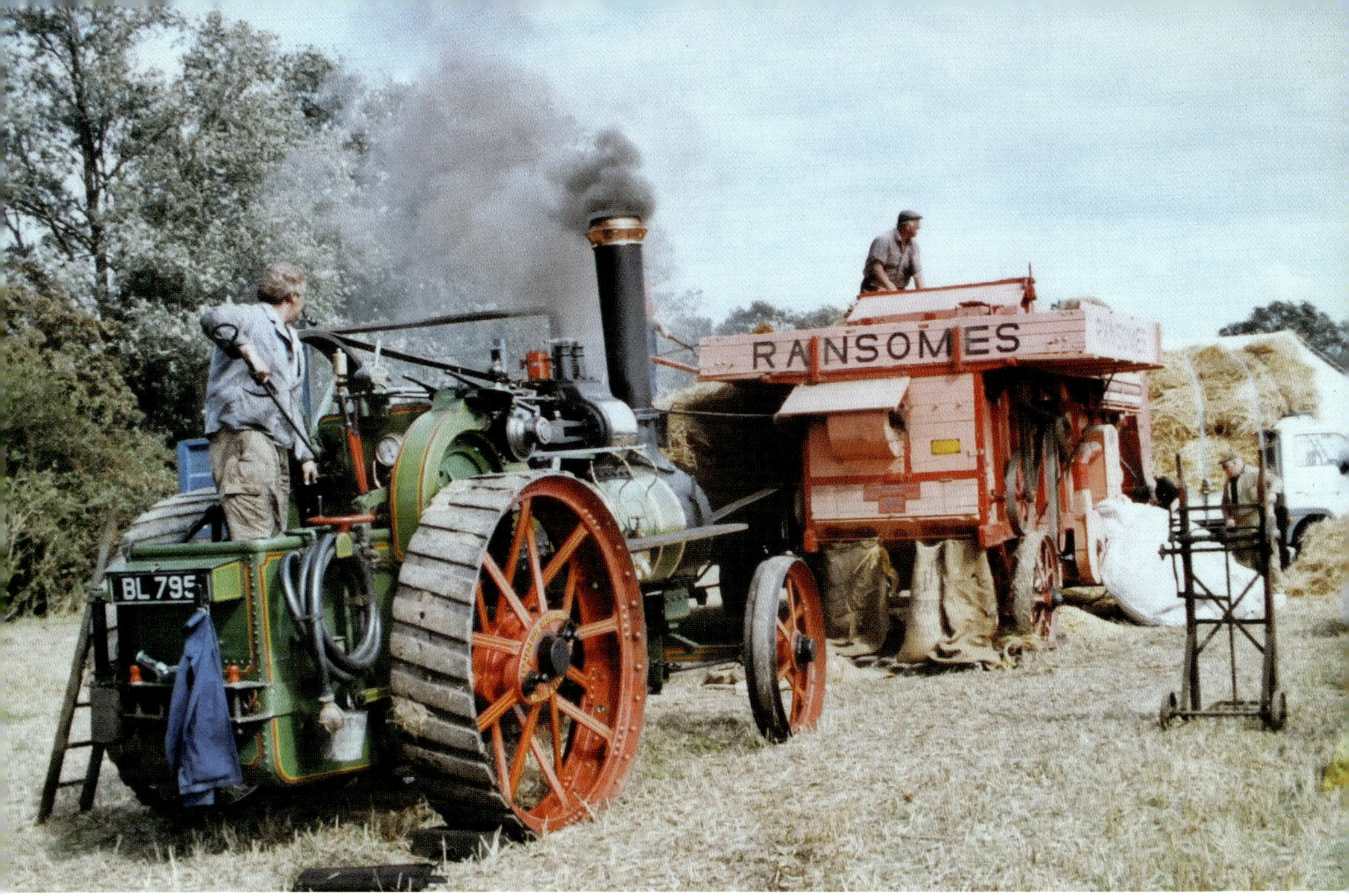
RANSOMES
BL 795

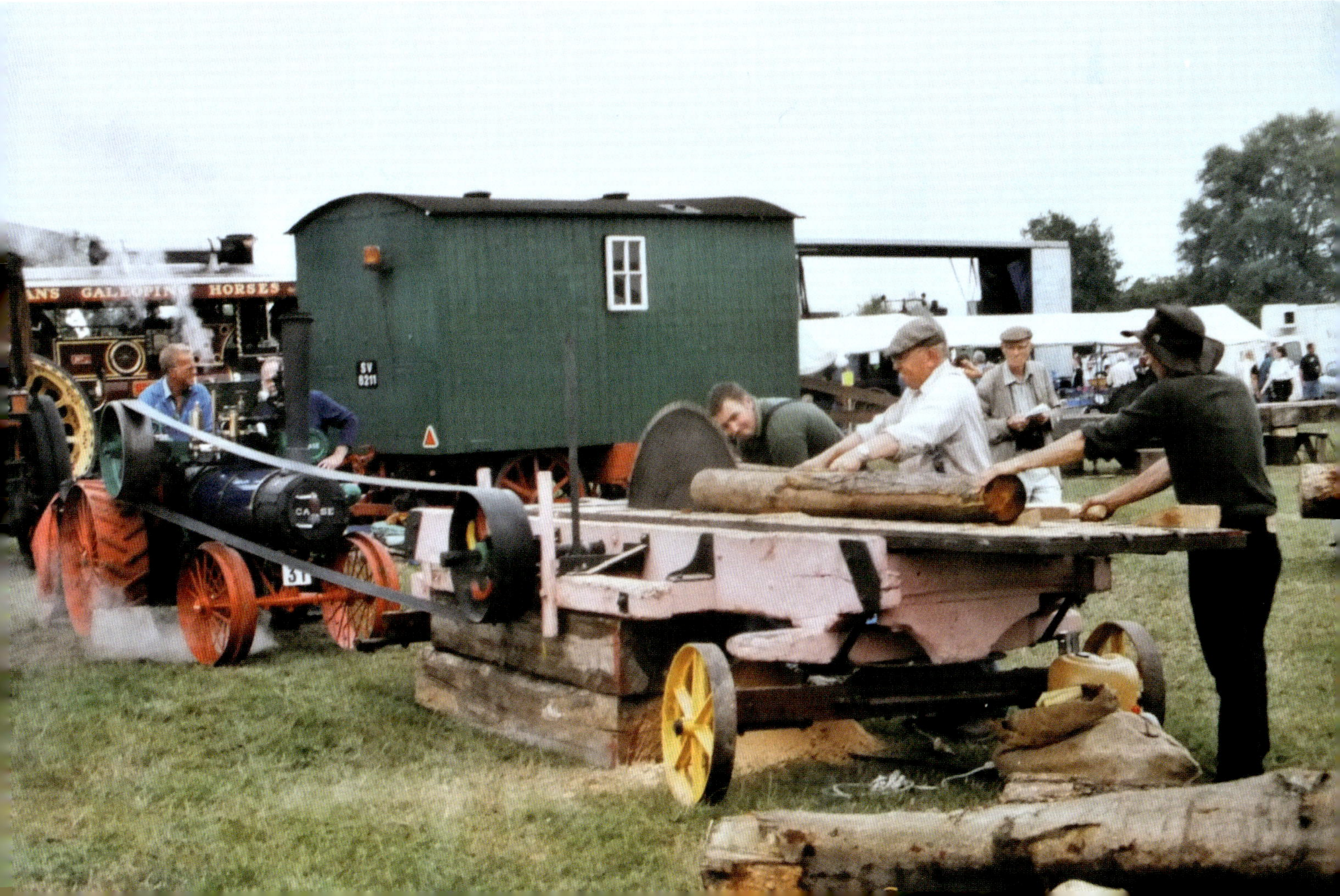
GALLOPING HORSES

Above: A roadmaking display area was gradually developed and enlarged over the years. Here in 2006 is Ruston & Hornsby roller No. 122338 *Endeavour* of 1924. Note the wooden tripod stands used in earlier days to fence off the road works, before the advent of cones and orange plastic sheeting; and the cart-mounted stone crusher to break down the stones into gravel for the rollers to roll into the road surface.

Opposite above: The oldest Fowler engine in the Register is No, 1050, a ploughing engine built as early as 1868. This was rebuilt by Thomas Wood, Crockenhill, Kent c.1911-2 with a Burrell single crank compound cylinder block. Still based in Kent, this was KE 2319 performing in 2008. In the past this had been incorrectly reported as No. 2620 and as registered KE 2322.

Opposite below: A highlight of the 2009 rally was the line-up of five Ruston steam rollers, plus one that had been converted to diesel. The diesel conversion is No. 52694 *Nuffrush* which is fitted with a Nuffield tractor engine. Two of the rollers XM 6374 and XM 6373 were new to Henry Woodham & Son, Catford, as was Wallis & Steevens roller OT 3317 also present.

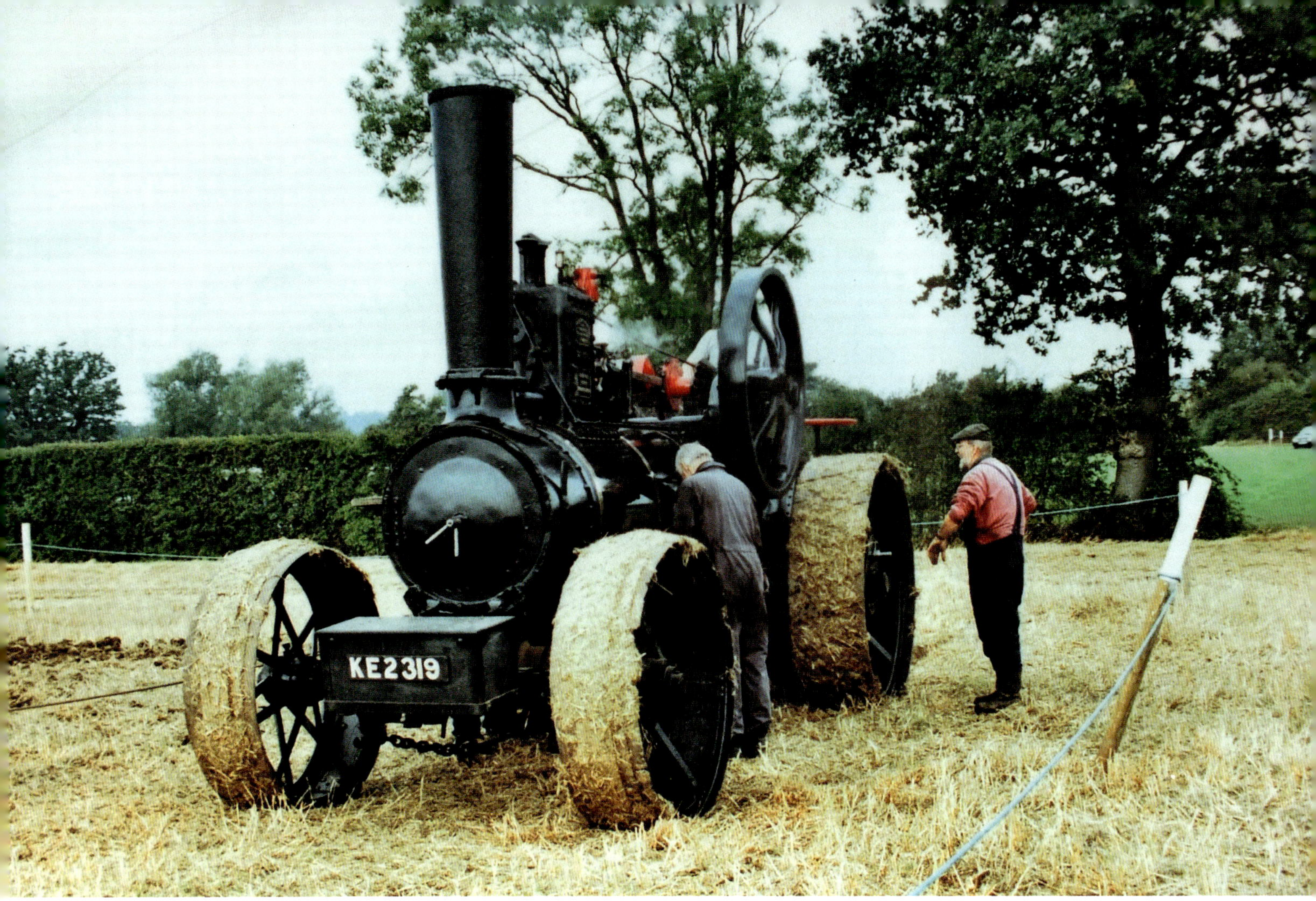
KE2319

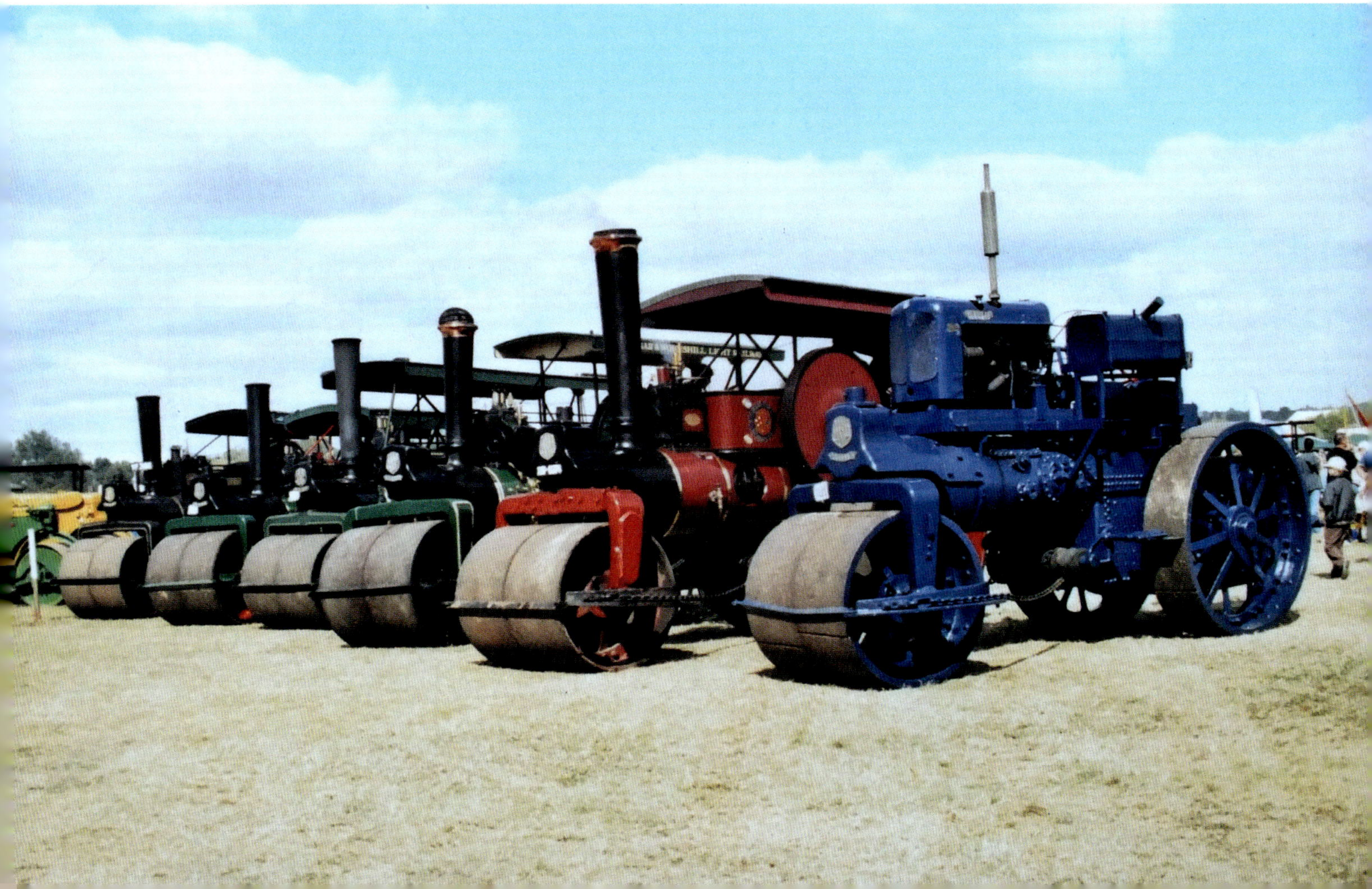

Above: After the Burrells from New Zealand came five McLarens in 2010. Nos 1266 and 1242 were at Barleylands in 2010.

Below: Also visiting from New Zealand was McLaren tractor No. 1817 of 1926 which is a conversion from a roller.

Above: Sentinel S4 wagon No.8827 was in Holland from 1981 to 1998 (see p. 92) Now back in the UK it had been repainted into the colours of the Gas Light & Coke Co. when it featured at Barleylands in 2010.

PRESTON RALLY & COUNTRY FAIR
2003–2014

Preston Services at Preston, near Wingham, Kent are dealers specialising in buying and selling steam engines and spare parts. They originally held Open Days just after Christmas which lasted from at least 2002 until 2013. A summer steam rally was first held on 28-9 June 2003 which then became an annual event until 2014. On display each year was a fine selection of the owner's extensive collection of engines, a range of his dealer stock, and various visiting engines.

An unusual feature here, that was first tried in 2008, was that the attending engines were supplied with locally grown and selected hardwood rather than coal. It was claimed that this made the Preston Rally the world's first 'Green' rally, running on renewable fuels, reducing carbon emissions and chimney smoke (and also fewer black smuts on clothes!)

ANNUAL RALLIES THAT HAVE CEASED • 79

Above: In 2008 was Ransomes, Sims & Jefferies wagon No. 34270. This was rebuilt from parts repatriated from Australia with the construction of a new chassis and running gear to create the only wagon by this make in the Traction Engine Register.

Below: Of distinctly American appearance was this engine built by the Frick Company of Waynesboro, Pennsylvania, amongst the dealer stock in 2008. This is No. 21918 of c.1921.

Above: A pair of the miniatures present get their turn around the ring in 2008.

Below: More rarities on view in 2009. This is Henschel No. 5063 *Helga*. Built as recently as 1953, this was the last steam roller to be built in Germany and was imported to the UK in 2004.

Above: There are only five Davey Paxman traction engines in preservation in the UK, so you don't see one very often. One that is Kent based and registered is No. 19412 of 1916, registered KE 2700 which was on hand in 2009.

Opposite above: Preston Services have been responsible for the import of many engines, especially portables from South America, many of which had not been identified at the time the 2016 Traction Engine Register was compiled. This view shows just some of the portables that were on the premises in 2009.

Opposite below: A wagon that has been with the founder of Preston Services, Mr List-Brain, for many years is Atkinson 6 ton 'Colonial' type No.72 of 1918. This was imported from Australia in 1976 and is the only known survivor of some 540 wagons built by Atkinson from 1916 onwards.

The ATKINSON
SAND &
GRAVEL
TEL.
322
12
LAKE
GOLDSMIT
MINING PTY. LT
AUSTRALI
CK 209

Above: Burrell No. 2342 *Vanguard* is a showman's road locomotive dating from 1900. This was moving through the site in 2009.

Opposite above: Fowler K7 ploughing engine No. 13310 dates from 1914. This bears the name *The Steam Sapper* and was brought back from Germany in 2005.

Opposite below: A massive beast is the German built Kemna/Ottermeyer 16nhp ploughing engine *Zeven*. The identity of this is recorded as 'uncertain', possibly built c.1918 but with a replacement boiler dating from as late as 1958; it drew a plough which created a furrow 9ft deep and 8ft wide for reclaiming areas of peat moorland in central Germany. Seen here in 2010, this is now owned and has been restored by Robert Jardine of Doddington, Kent.

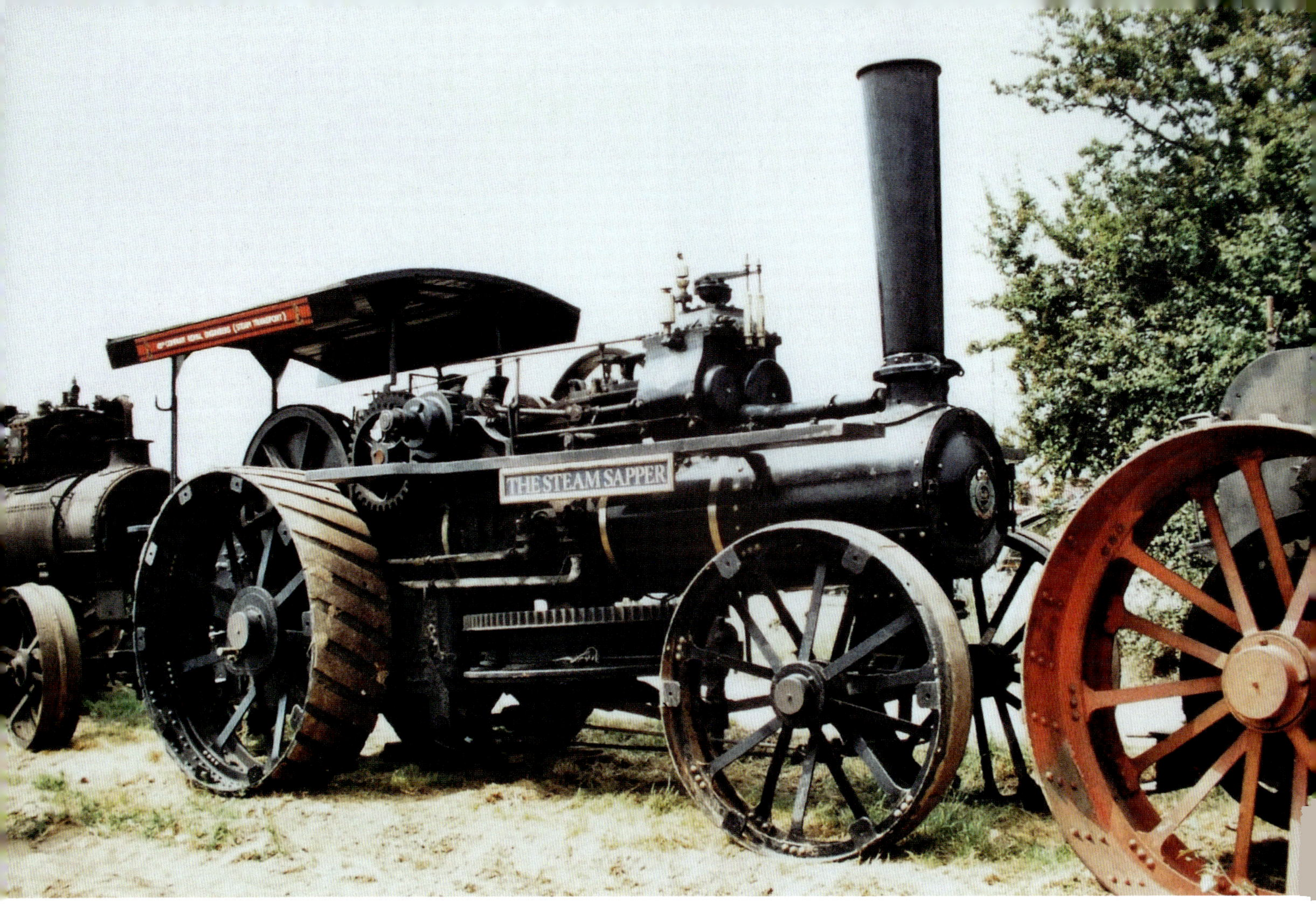
THE STEAM SAPPER

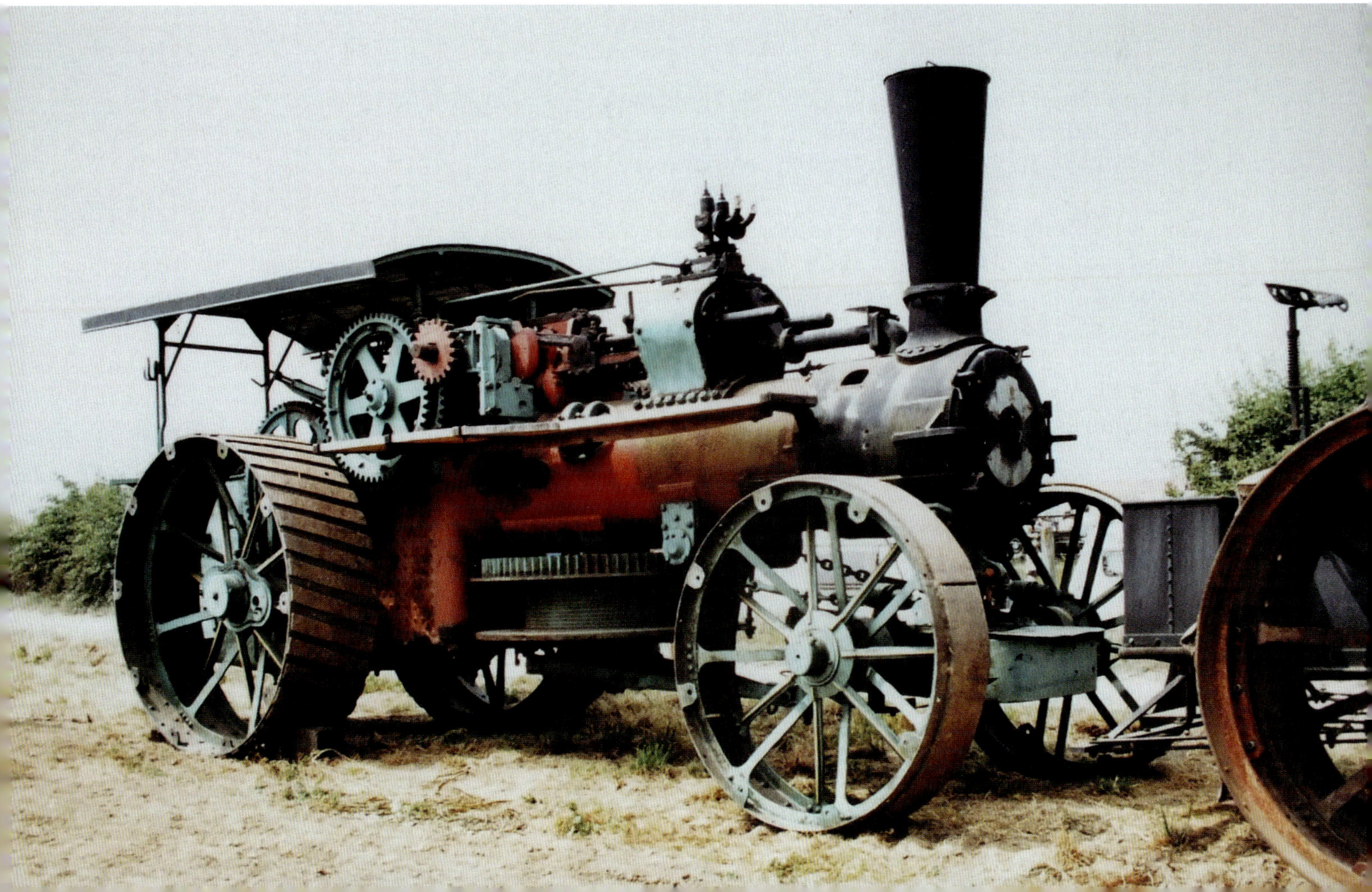

Above: With Preston church in the background, this is Ransomes, Sims & Jefferies No. 25781 in 2011. Dating from 1913 this massive machine is a direct ploughing engine which was brought back from Argentina in 2002. This has since been sold on and fully restored. (see p. 124)

Opposite above: Ploughing could take place at Preston. Looking more like a railway locomotive on road wheels is this American built Avery engine of 1914 which has undermounted motion. This was demonstrating direct ploughing, pulling the plough behind it at the 2011 show. It would have been quite a job to turn it round at the end of a furrow.

Opposite below: Another US import. This roller was manufactured by the Kelly-Springfield Road Roller Co. of Springfield, Ohio and is the only engine of this make in the UK.

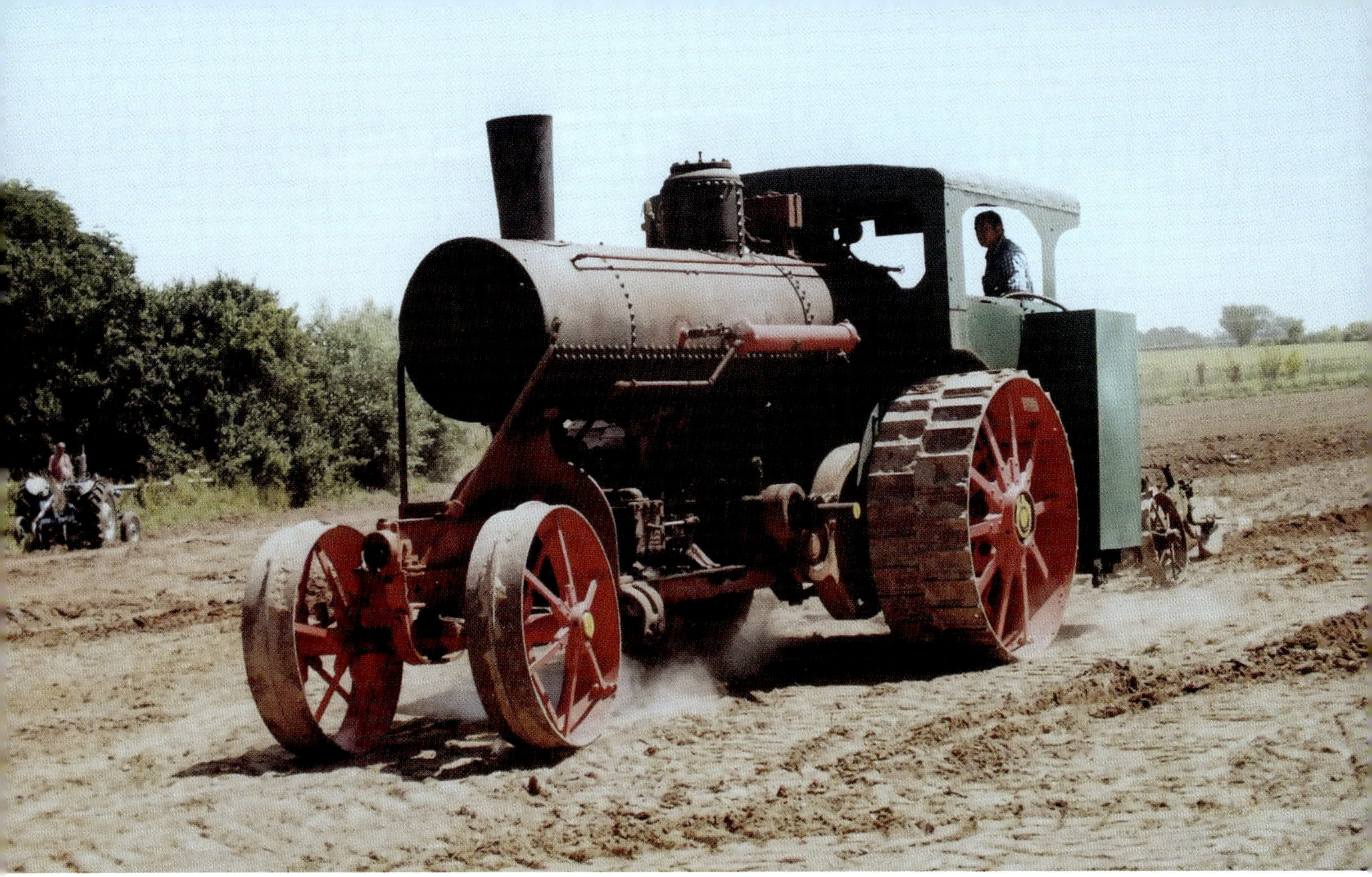

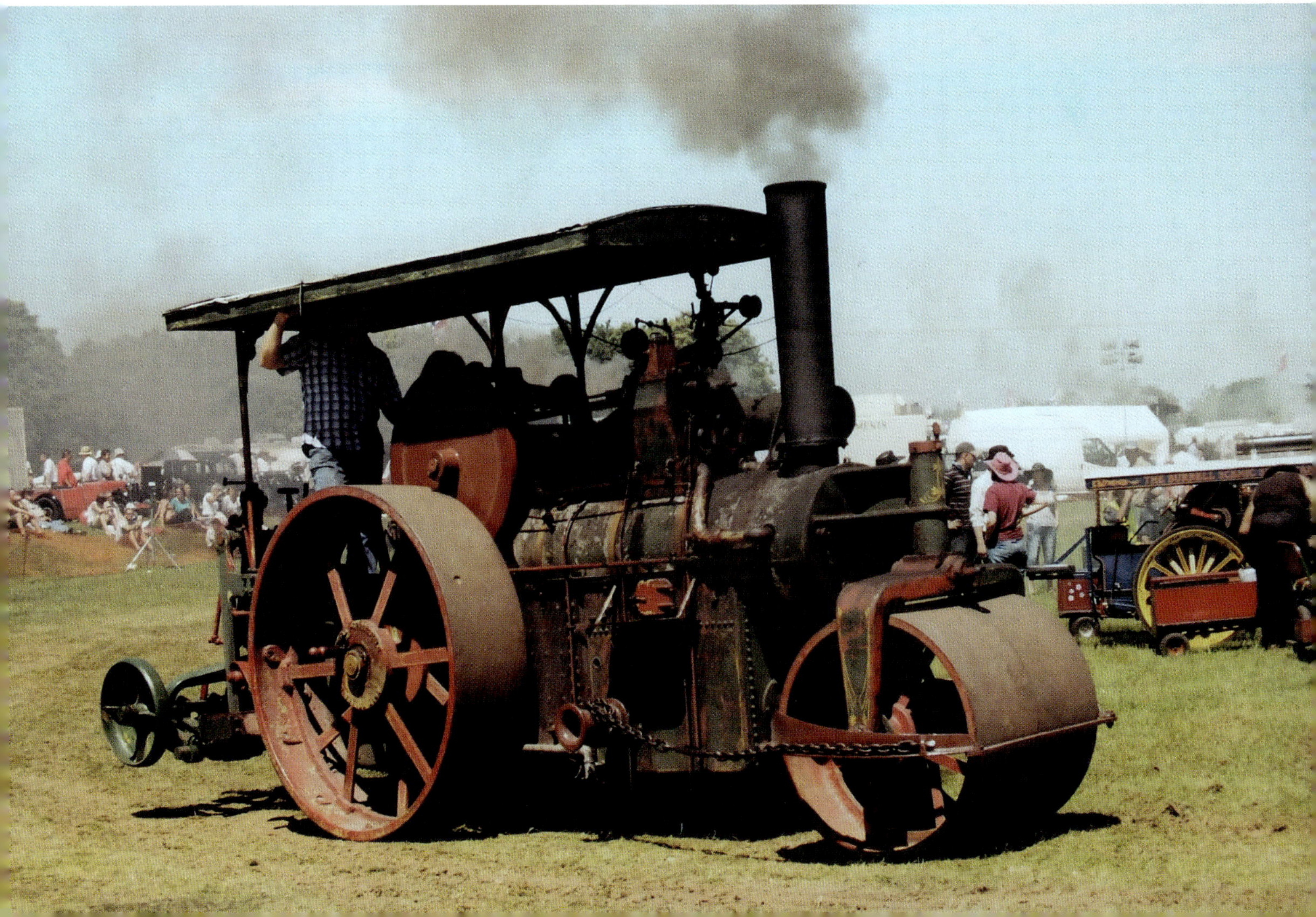

ANNUAL RALLIES ONGOING

HCVS LONDON TO BRIGHTON RUN 1962–DATE

The London to Brighton Historical Commercial Vehicles Run, organised by the Historic Commercial Vehicle Society (originally Club) has taken place annually since 1962 (except in 2020-1, cancelled due to Covid-19). This was the first of a number of annual road runs to be organised by the HCVS. While not a traction engine rally, steam powered commercial vehicles have always been entitled to take part and this gives them a rare opportunity to participate in a long-distance public road run. Inevitably it is the wagons, tractors and road locomotives that tend to feature, as these were the types of traction engine designed for such journeys. However, at least two rollers have completed the run, and a ploughing engine has attempted it.

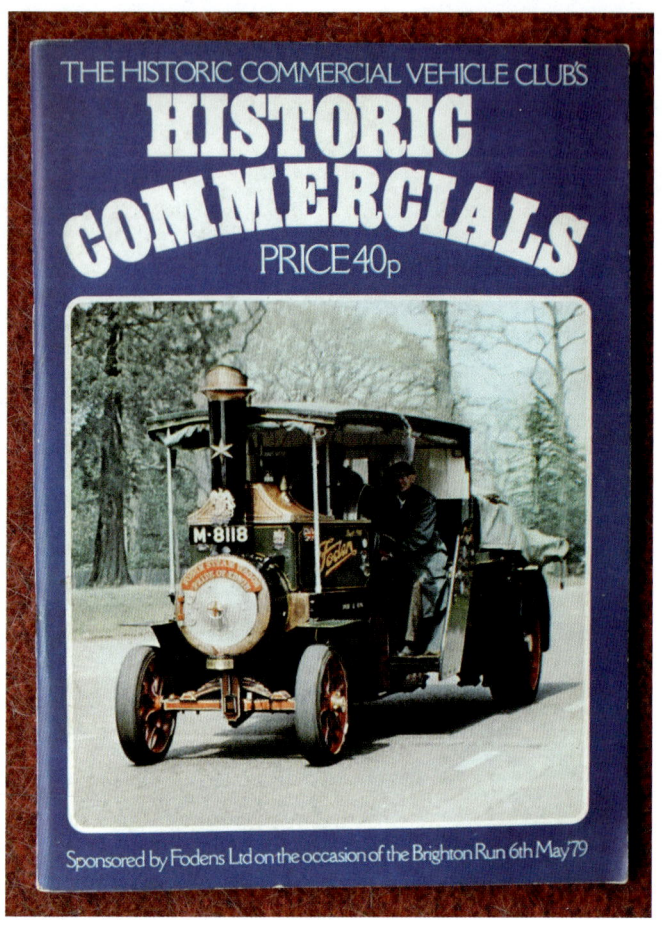

Opposite above: Steam wagons have been a feature at Brighton most years, and in 1972 we see 1931 Sentinel DG4P type No. 8571, KF 6482. Sentinel were one of the main makers of steam wagons, their models featuring vertical boilers and underfloor mounted engines. In the 1940s, they switched to diesel lorries and buses, but again featuring underfloor mounted engines. No. 8571 was supplied new to Samuel Banner & Co., Bootle, passing to Paul Bros, Birkenhead in 1936, with whom it would remain in service until 1949 when it passed into preservation.

Opposite below: At the end of the 1976 Run, onlookers watch as 1929 Foden steam wagon RO 6330 is prepared for winching onto a low loader trailer for its journey home. Foden were the other main maker of steam wagons beside Sentinel, but unlike Sentinel they specialised in the 'overtype' design, with an in-line traction engine style boiler and engine mounted above. Final drive to the axle was by chain. Originally bought by Hertfordshire flour millers, this wagon was later sold to Taroads Ltd and converted to a tar spraying unit in which form it lasted to the mid-1950s.

258
PAUL BROS LTD
HOMEPRIDE FLOUR MILLS
BIRKENHEAD
6

RO 6330
TREGUNTER
STEAM HAULAG
LONDON

Above: In 1978, Foden were the sponsors of the Run, but this entrant was from their main rivals Sentinel. BRF 200 is a 1933 built S6 model (shaft drive, 6 wheels) with a tipper body used for carrying roadstone. By the 1930s, Foden also introduced an undertype steam wagon, but with road tax discriminating against steam vehicles, and the more economical diesel engine replacing petrol, they would soon after cease making steam vehicles in favour of the internal combustion engine. BRF 200 was last used commercially during the 1956 Suez crisis when there was a shortage of oil fuel.

Opposite above: Taken in Battersea Park at the Wheels of Yesterday Rally in 1982, this 1926 Sentinel Super tractor and trailer had been a Brighton entrant the previous day. Restoration was completed between 1967-71. The Wheels of Yesterday Rally began in 1980 and would switch to Crystal Palace when the departure point for the Brighton Run moved there in 1989 but it declined and ceased in 1995.

Opposite below: Representing the steam entries in 1985, this superbly presented Foden wagon dates from 1931, although it was still on solid tyres. This was the same year that the company made their first diesel lorry. It was new to Derbyshire Council as a tipper and had been entered on the run in 1972 fitted with a tipper body (see p. 26). By 1977 it was carrying a tar tanker body (see p. 54). Foden No.13848 is now resident in the Isle of Man.

JOHN R. HARRISON
Contractor
GARLANDS
STEEPLE BUMPSTEAD
NT 8347

RADCLIFFE
100 MILE JOURNEYS COMPLETED IN ONE DAY
CARRIERS
STORERS
RADCLIFFE & SON
LONDON
MANCHESTER
THEATRE CARTAGE
GENERAL HAULAGE
RB-3525

Above: In 1987 this 1933 Sentinel S4 steam wagon No. 8827 was entered by its preservationists in Holland. It has since been repatriated and regained its original UK registration, UJ 2225.

Opposite above: Also in 1987, an ambitious attempt was to enter a 1917 Fowler ploughing engine on the Run. Unfortunately, NO 5 did not manage to complete the journey under its own power and arrived at Brighton on a low loader.

Opposite below: Burrell showman's road locomotive No. 3949 *Princess Mary* was new in 1923 to Billy Nichols of Forest Gate, London, and later sold to Charles Presland of Tilbury. Now preserved in Dorset, in 1988 it completed the Run hauling a pair of showman's trailers just as it would have hauled during its working life. Also completing the Run was Fowler showman's road locomotive No. 15653 *Renown*. They set off from Battersea Park at 2.10am. The trailers were provided by Harris's Amusements of Sussex.

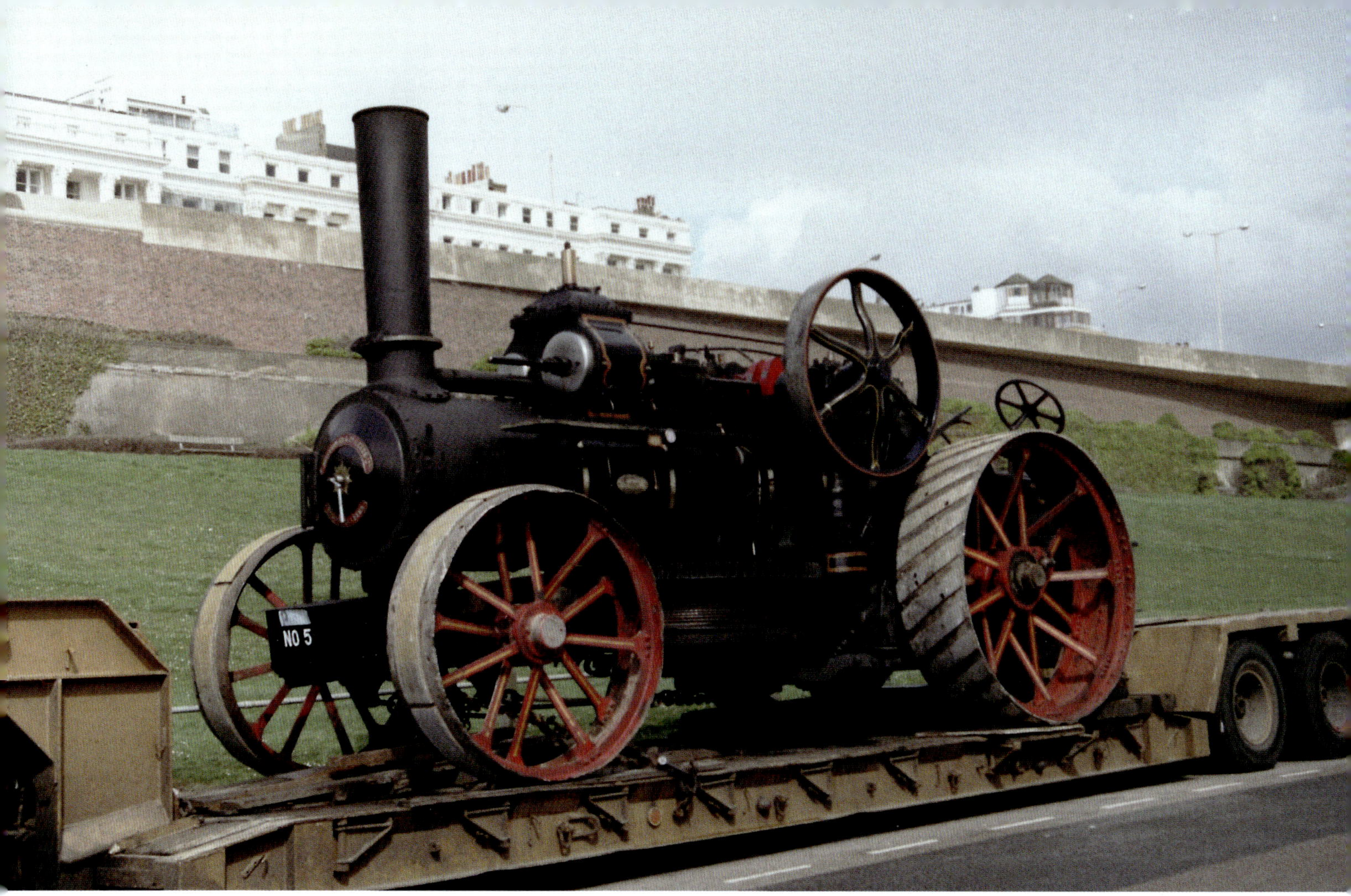
NO 5

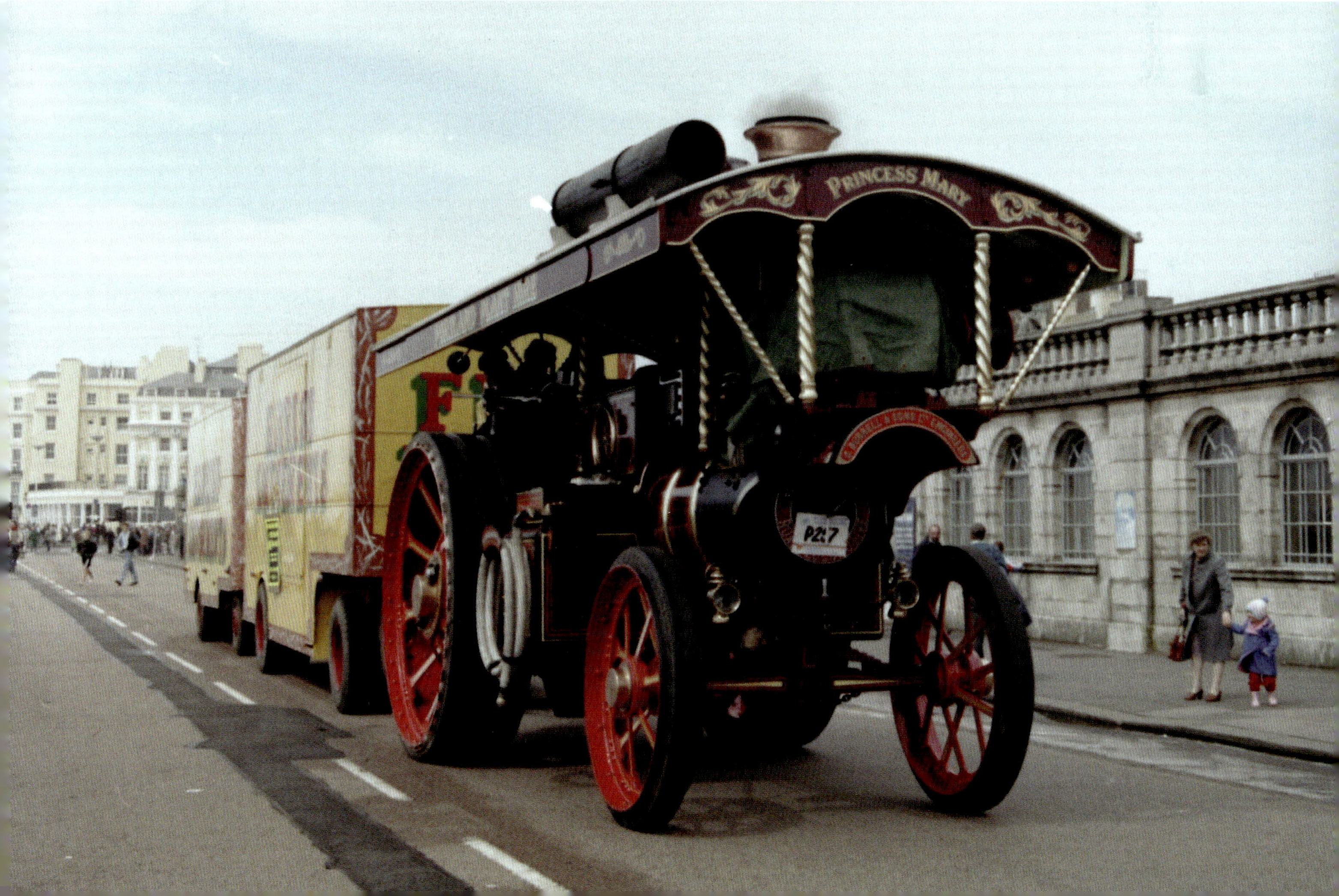
PRINCESS MARY

Above: Foden wagon No. 3510, M 4673 has an interesting story. Built in 1913 for J.T. Lawton Ltd of Manchester, it was purchased in 1928 by Henry Ford for his museum at Dearborn, Michigan. It would remain static there until repatriated in 1980. When seen on the Run in 1998 it still carried its original paintwork, its long period in the museum having prevented its deterioration.

Opposite above: It is rare for an agricultural traction engine to make the Run. Slow speed and low water capacity are obvious logistical problems. In 2004, Marshall No. 78667 of 1925 completed the Run hauling two trailers. This had been acquired by the owner in 1997 (when he was aged 19) as a kit of parts and rebuilt over the following six years.

Opposite below: Burrell 4nhp Gold Medal tractor No. 3442 of 1913 completes the Run with a pantechnicon trailer in tow in 2009. This is just the sort of work these useful tractors would have performed during their working lives, although this example was used for timber haulage, winching and sawing with S.M. Price, Craven Arms. It has been with the Searle family of Horsham since 1986.

HO 6360

FIREPROOF DEPOSITORIES
L.C. LAMERTON
PACKERS SHIPPERS & STORERS
EALING, LONDON
EXPERIENCED PACKERS
AH 0119

Above: The Searle family from Horsham tackled the Run in style in 2011 with this impressive feat of road haulage. Leading the line is Fowler crane road locomotive PB 9687 *His Majesty* (with crane removed). Behind this is McLaren 1919 10nhp road locomotive WF 1864 *Boadicea*, and on the rear of the trailer is 1901 Fowler crane road locomotive SG 4713 *The Great North*. The idea was to recreate a photograph from working days showing *His Majesty* on the rear of a similar move. With a trailer loaned by Martin Oliver and a Fuller transformer that was due to be scrapped all the requirements were in place and a VSO (vehicle special order) was obtained from the police authorities for the journey. Crystal Palace was left at around 2am and the journey raised money for the Teenage Cancer Trust. The next generation of the Searle family (known as 'The Creche') repeated the feat for the 2022 Run.

Opposite above: In 2015, the celebration of the centenary of the First World War was well underway and several of the Run entrants were vehicles from this period. McLaren 1912 10nhp road locomotive No. 1332 was entered in War Department livery hauling a replica howitzer and gun carriage. Although this particular engine did not see war service, having been exported to Australia when new, other McLaren road locomotives were used as gun tractors.

Opposite below: Aveling & Porter 1919 built roller BH 7167 *George* finally arrives at Brighton in the afternoon after an epic journey from Crystal Palace, London to celebrate its centenary year in 2019. We saw this roller earlier in the 1960s. (see p. 19)

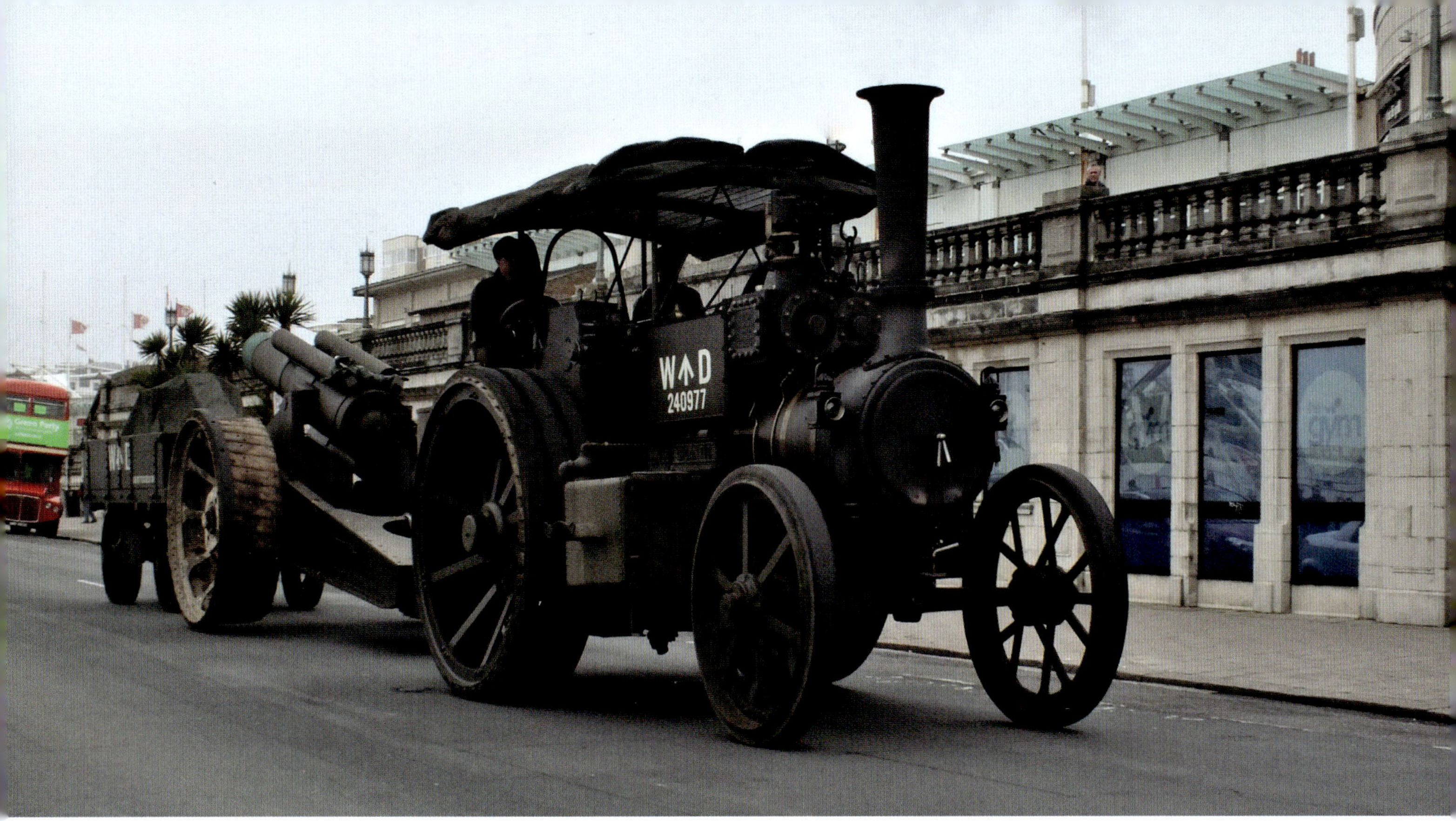
W↑D
240977

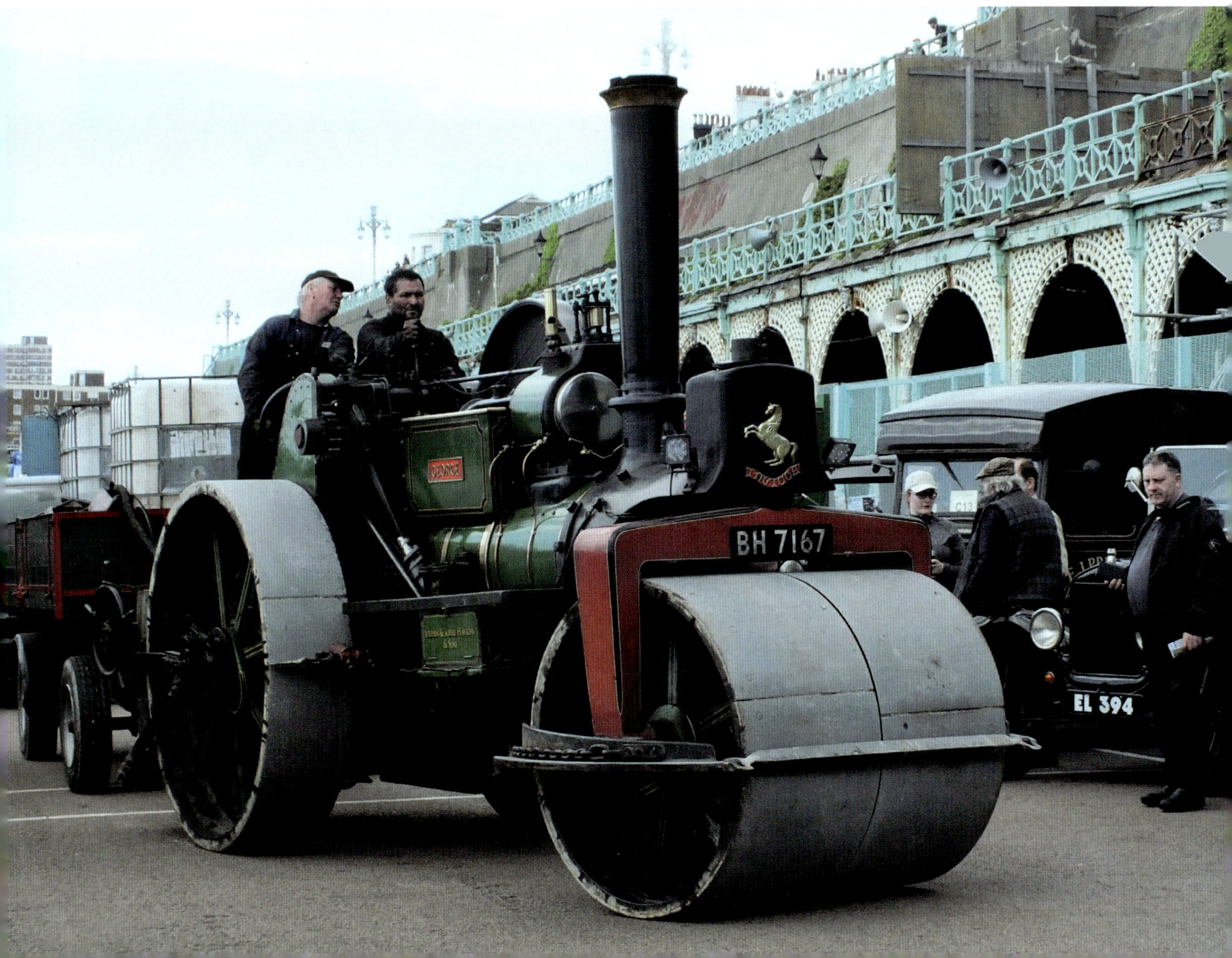
BH 7167
EL 394

BANBURY VINTAGE VEHICLE AND COUNTRY FAYRE 1969–DATE (NTET AUTHORISED EVENT)

Oxfordshire is where the traction engine rally movement started, so it is fitting to feature a rally from that county. The Banbury Steam Society, established in 1968, have been staging their annual rally since 1969 which takes place at Bloxham near Banbury. The 2004 event was advertised to feature fifty engines to celebrate the fiftieth anniversary of the NTET.

Below: Bloxham in 1992 and a Sentinel wagon that originally hailed from my local area of east London. The HV registration is from the County Borough of East Ham where E&A Shadrack Ltd. were coal merchants. Sentinel S4 No. 9016 dates from 1934.

Opposite above: A visit in 1995 and this the only surviving engine listed from the company of Gibbons & Robinson, No. 959, AY 9874 of 1891. The company came from Wantage, Oxfordshire so its appearance here was appropriate.

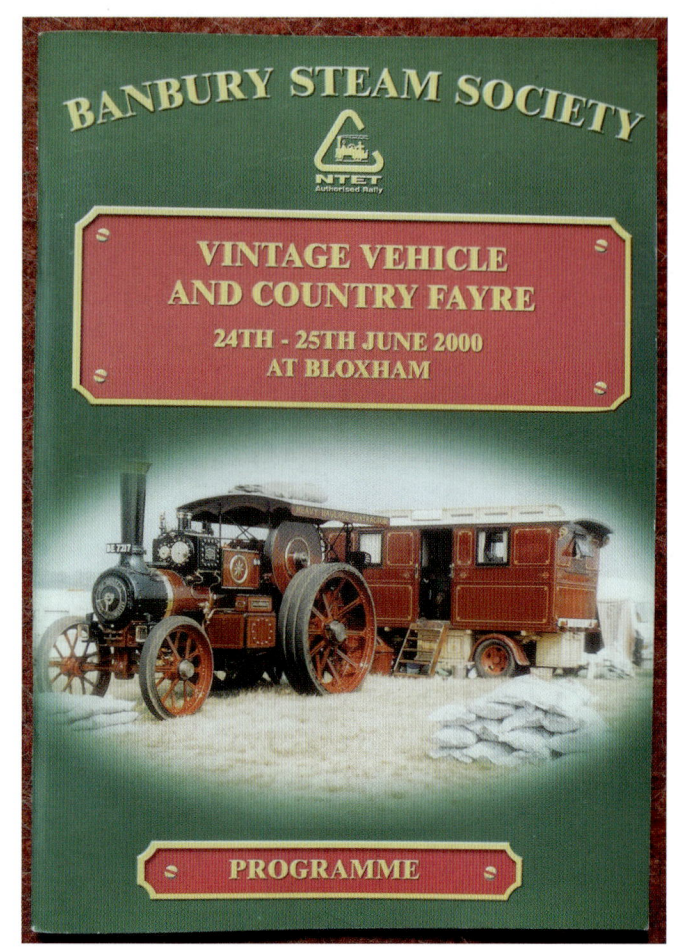

Right: In 2000, another rare old veteran was Hornsby 8nhp engine No. 6557 *Maggie* of 1889. This is one of only three Hornsby traction engines that survive in the UK dating from the period before 1918, when the company merged with Ruston, Proctor & Co. Ltd to become Ruston & Hornsby Ltd (although there are also several portable engines from this period). The other two Hornsby traction engines were repatriated from Australia. FL 2598 was rescued for preservation in 1963 and restored from 1977 onwards. It had then passed to the present owner from Oxfordshire in 2000.

Left: A company trading variously at different times as Kirby & Barrows, Barrows & Carmichael, Barrows & Stuart and Barrows & Co. Ltd were manufacturing in Banbury between c.1862 and 1919, so it was interesting to see one of their products at Bloxham in 2000. The board states that this portable engine was built by Barrows & Carmichael in 1867. Its early history is not known but it finished its working life with Thompsons of Buckden, Huntingdonshire who were agricultural engineers and contractors. The portable was purchased from a scrap yard in Bedford in the 1970s and restored.

Below: A locally-based Fowler BB1 ploughing engine is No. 15182 *Achilles* of 1918. This 16nhp engine is seen in 2013 waiting to enter the ring.

Opposite above: A smaller Fowler seen in 2015 is No. 8889, WR 6554 *Dreadnought*, a 5nhp tractor. This is actually a rebuild in preservation from a 1901 built roller. It was standing alongside a Stanley steam car.

Below: The lettering on 1929 Sentinel DG6 wagon No. 7966 is a credit to the art of the signwriter. Restoration was completed by Barry Cousins in 2009. This was the last of thirty-two Sentinels supplied to J. Lyons & Co. and the only DG6 model. It only worked until 1933, a victim of prohibitive speed legislation on solid tyred vehicles.

Above: At the 2016 rally, Savage centre engine No. 657 *Lydia Grace* was present, celebrating its 120th anniversary year.

Opposite above: Looking immaculate in 2017 was Fowler roller No. 21636 *Covenanter* of 1938. As the canopy lettering states, this was new to Ayr County Council but is now preserved locally in the Banbury area.

Opposite below: Making a first rally appearance in many years at the 2018 Golden Jubilee Banbury rally, although not in steam, was *Maude*, Foster No. 3642, PB 9832 of 1908.

AYR COUNTY COUNCIL
CROOKEDHOLME
ROAD REPAIR DEPOT
CS7859
LANG.

ROSES·UPTO·DATE·AMUSEMENTS
MAUDE
LINCOLN, ENGLAND
309 KTM

BEDFORDSHIRE STEAM & COUNTRY FAYRE 1957–DATE (NTET AUTHORISED EVENT)

The Bedford Steam Engine Preservation Society (BSEPS) was formed in February 1956. The founder and first Secretary was John Crawley, who had collected a number of engines and arranged 'steam ups' at local pubs. The first rally was held at Woburn Park on 5 August 1957. This site was used again until 1967, although there were no rallies in 1959 or 1962. In 1962, the Woburn Park site was used in the filming of *The Iron Maiden*, which featured engines owned by John and other Society members. The 1968 rally was held at Houghton Conquest. The Bedfordshire rally, by now a two-day event in mid-September, moved to Roxton Park where it was held from 1969 until 1993. The 1994-5 events were at Shefford. A new home then became established at Old Warden Park, home of the Shuttleworth Trust where it was to remain until 2018. During this time, there were a number of special themes celebrated, including Clayton & Shuttleworth products in 2000, steam fire engines in 2001 and 'Made in Bedford' in 2006. A 'play pen'

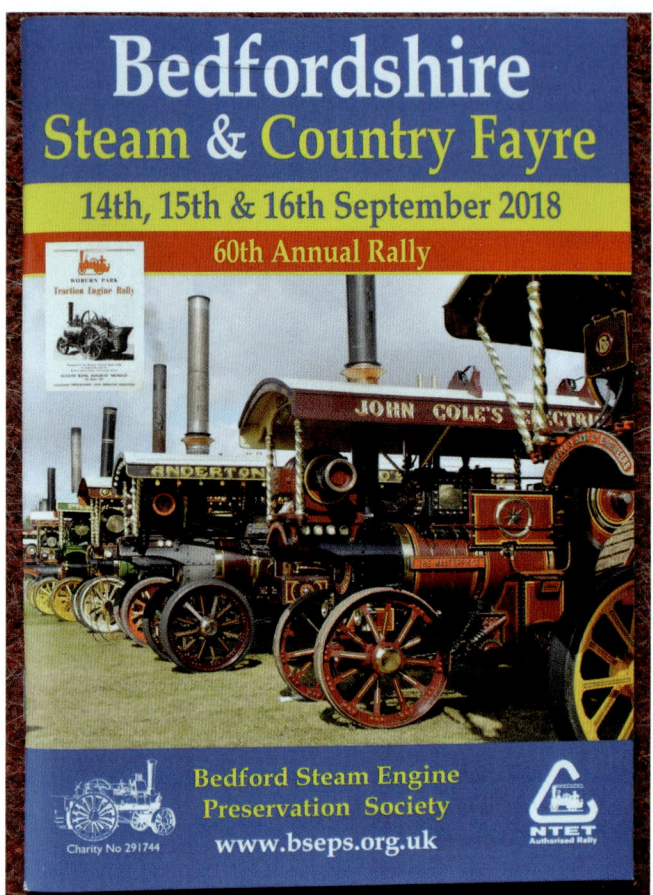

was added in 2014. At the 60th anniversary event in 2018 there were twelve engines booked that attended the first rally in 1957. There was no rally held in 2019 as the site was unavailable, and unfortunately the 2020 and 2021 Fayres which were to have been held at Turvey were cancelled, as were most other events because of the Coronavirus pandemic. The Bedfordshire rally ranks second only to the Great Dorset as having the highest number of full-sized engines regularly entered.

Opposite above: Appearing at the 1992 Roxton Park rally, this strange contraption was designed by John Kilgour and constructed in 1984 by the Silsoe Agricultural College.

Opposite below: At Shefford in 1994, this Marshall engine proudly displays the maker's name on the cylinder cover. YA 8855, No. 76963 was constructed in 1923 as a compound roller but was rebuilt to this form in the 1970s.

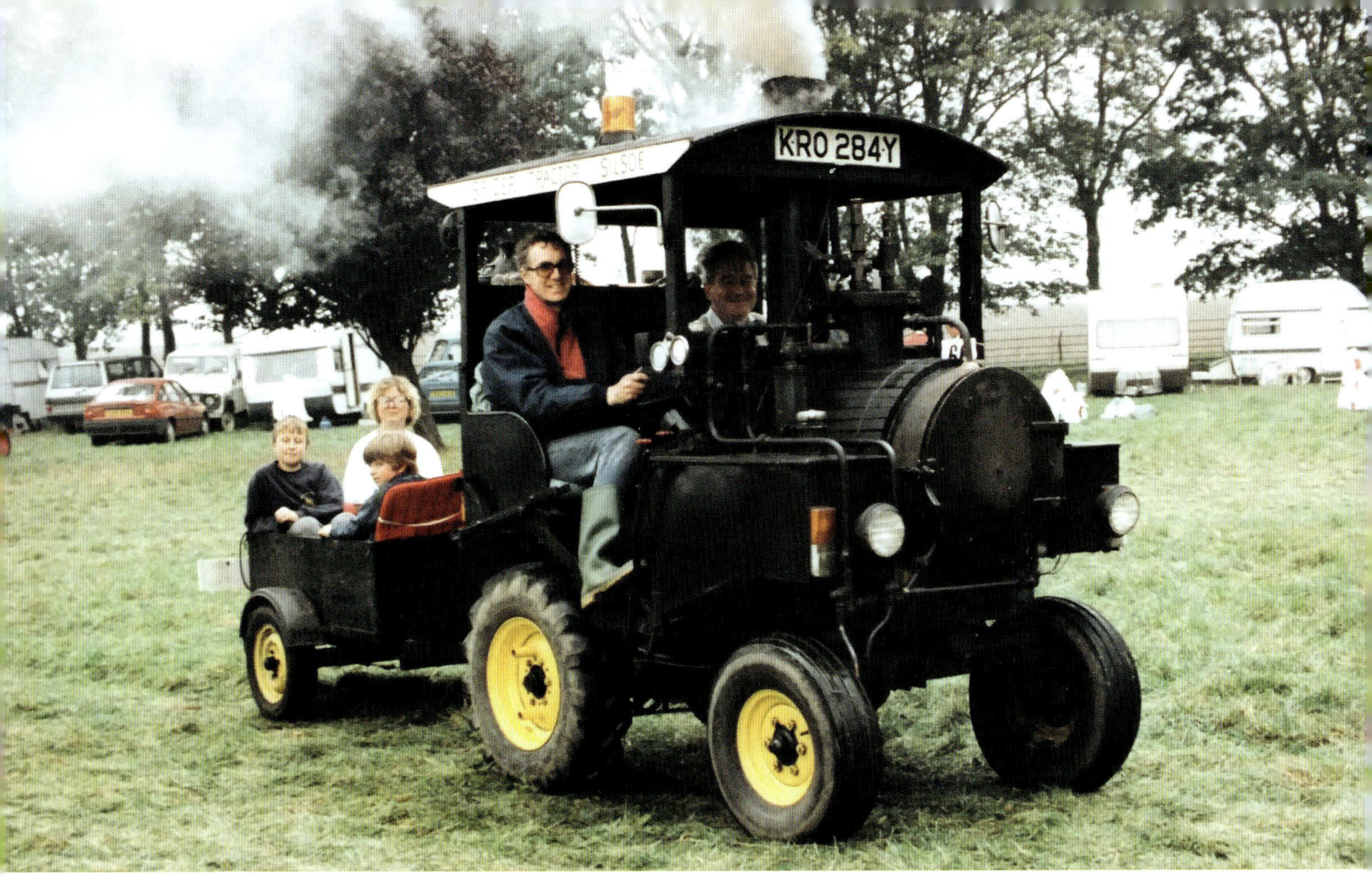
KRO 284Y

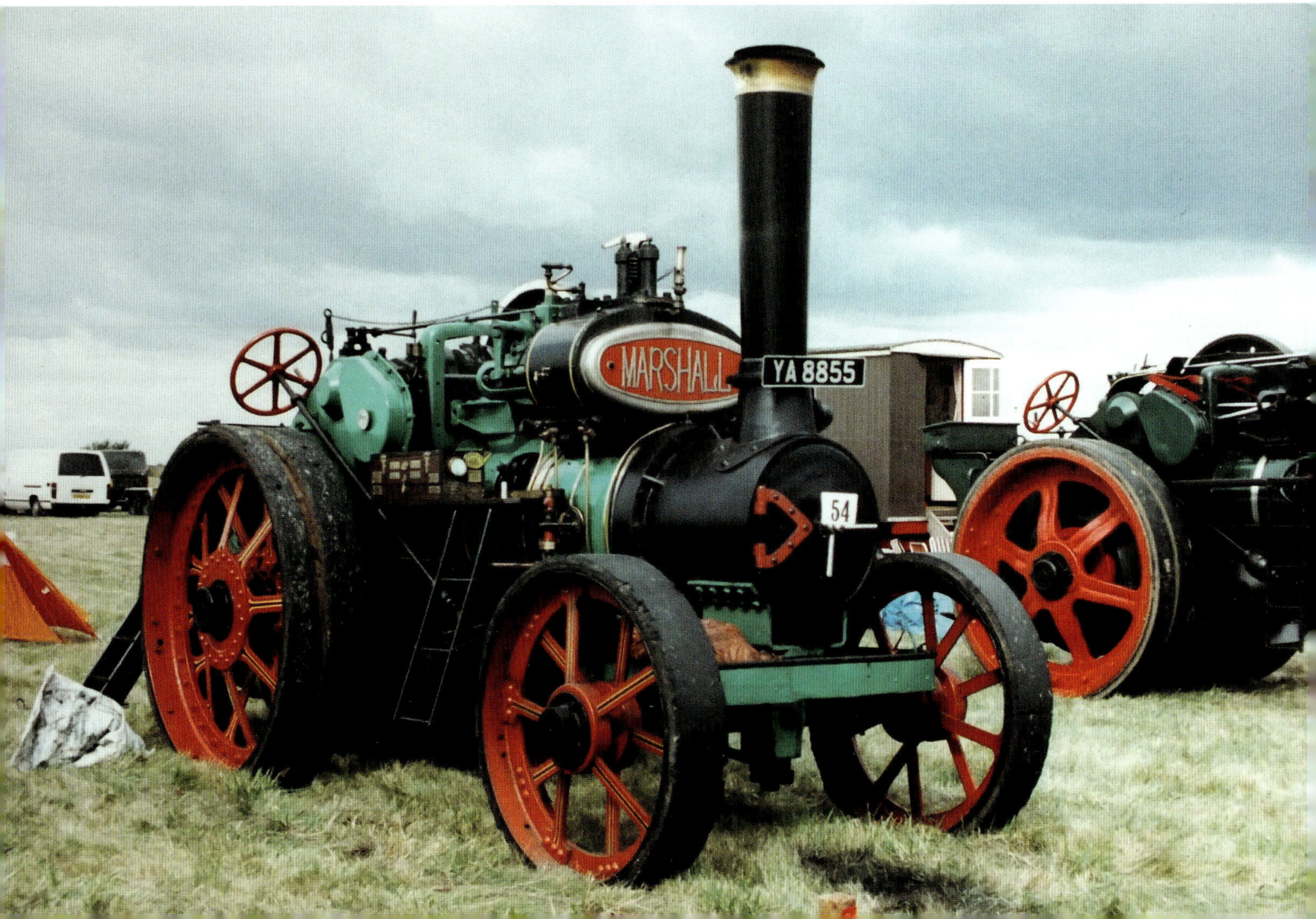
MARSHALL
YA 8855
54

Above: Passing the line-up of engines at Shefford in 1994 is one of a series of six 6in scale miniature Foden wagons built by the Maskell family in their workshops from 1992-2000. L59 GNM was built in 1993 and is a replica of the Foden Brass Band Bus.

Opposite above: In the ring at Shefford in 1994 is Clayton & Shuttleworth No.47015, BD 5483 *Dorothy*. A locally based engine from Bedfordshire, this 7nhp machine dates from 1914 and so would have been eighty years old at the time.

Opposite below: Horse power and steam power! Two Shand-Mason fire engines, dating from 1909 and 1876 pose in the ring at Old Warden in 1997.

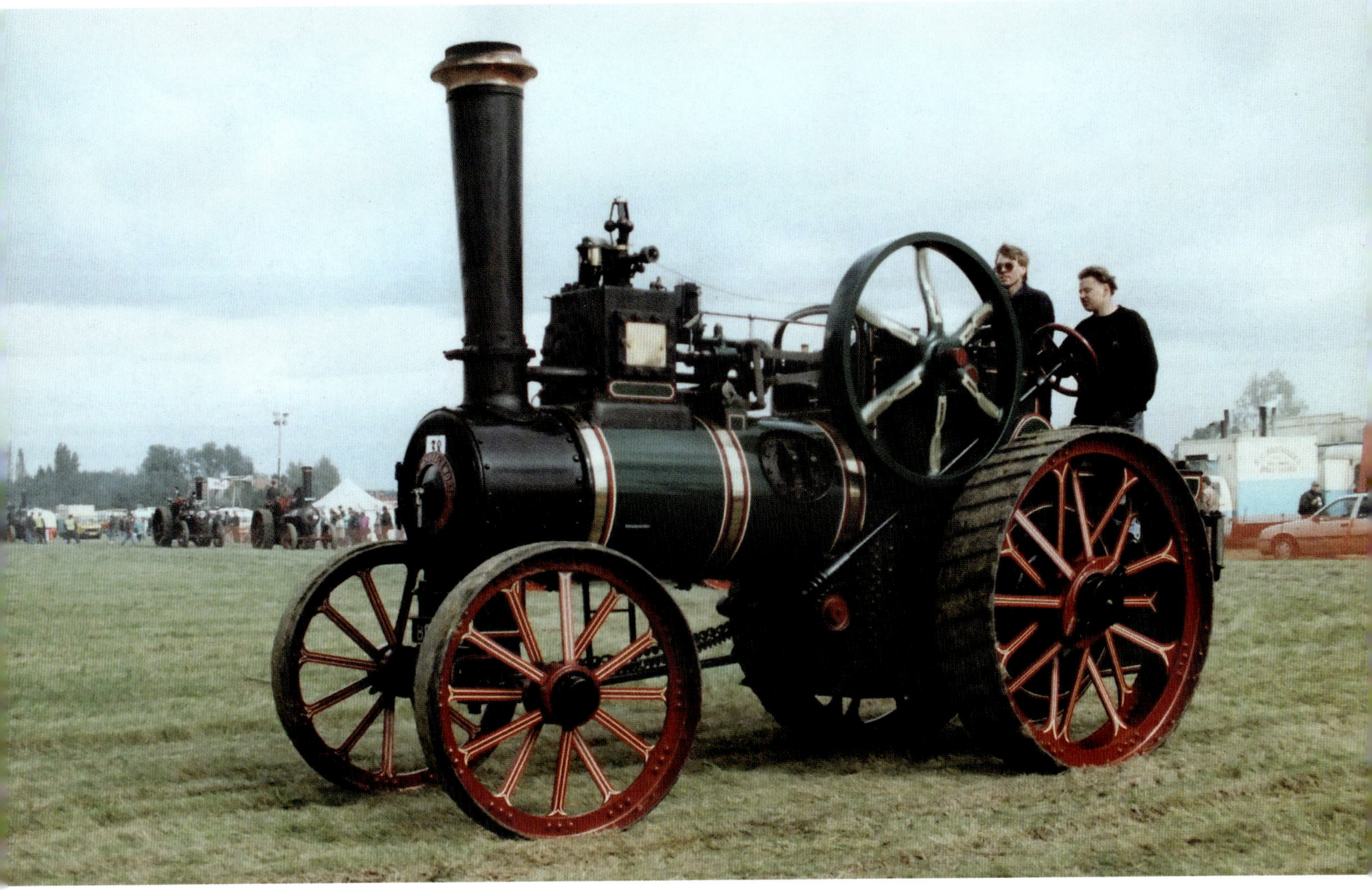

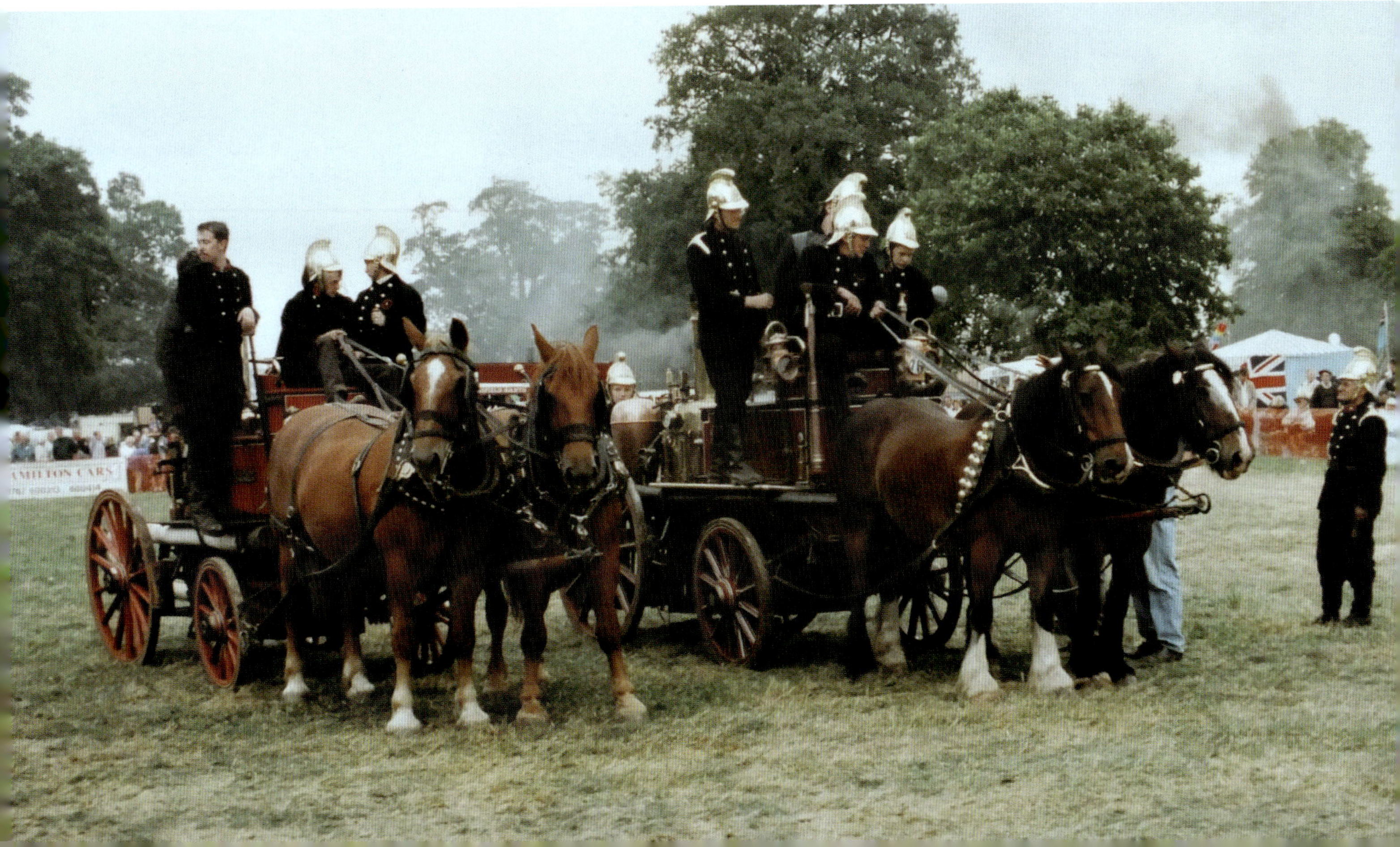
AMILTON CARS

Above: The 1876 Shand-Mason fire engine No. 2017, lettered for Sedgwicks Watford Brewery is a regular rally attender in the area. In 1998 it was paraded around the ring not by horses but towed by 1922 Ruston & Hornsby traction engine DO 2953 *Hildary*.

Below: Also in the ring in 1998 is Aveling & Porter tandem roller NM 291. Dating from 1908 this was the first steam roller bought by Luton Corporation. It is now kept at the Stockwood Discovery Centre in Luton.

Above: In steam but not yet fully restored, Howard ploughing engine No.110 was a visitor in 1998. J&F. Howard were a company from Bedford from which only this and one traction engine have survived.

Below: The Farmers Foundry Co Ltd, St. Andrews Works, Great Ryburgh, Norfolk was a small local company. Two portable engines made by the works survive. No. 36 of 1910 is now part of the Saunders collection and here it is driving a threshing machine at the 1999 show.

Above: In 2000, Clayton & Shuttleworth wagon FE 3344 *The Fenland Princess* waits to enter the ring. This dates from 1920. It spent much of its working life as a tar sprayer, eventually ending up in Hardwick's scrapyard, Ewell in a derelict condition.

Opposite above: I know this is not a traction engine, but you can hardly mention the Bedfordshire Steam & Country Fayre without mentioning the Saunders family collection from Stotfold. The collection was started by Reg Saunders in 1961 when he purchased Fowler roller No. 17560. He had previously purchased this 1930 Dennis GL truck and converted it to a mobile fish and chip van which had a coal-fired range. This was taken to early steam rallies and later replaced by a succession of converted buses. The Dennis, no longer used for frying, was retained and was displayed at the 2014 event at Old Warden.

Opposite below: The collection has been maintained and expanded by Reg's sons John and Ted. Pride of the Saunders collection is Burrell No. 4000 *Ex-Mayor* which was once owned by Sir William McAlpine (see p. 24). Here it stands in the fairground area at the 2003 Fayre, with the carousel rides and ex-Harry Lee Steam Yacht ride visible in the background. The engines in the collection are rallied widely and the showman's road locomotives are often accompanied by their 98-key Gavioli Scenic organ and the stage show.

SUPER FISH & CHIP BAR
R.C. SAUNDERS
SUPER FISH & CHIP BAR
DENNIS

WT 8606
G. T. TUBY & SONS
AMUSEMENT CATERER STONEBRIDGE PARK, MIDX

Above: As the Old Warden site is an agricultural college, the fields are cultivated, and depending on what has been planted, ploughing demonstrations have sometimes been possible. 2003 was one such year, and in fine sunshine, Fowler K7 type No.14257, KE 2494 *Linkey* is one of a pair on the plough.

Below: Ploughing was taking part on a field separated from the main site, so there was nothing in the background here to belie the impression that this could have been taken in the days of working steam.

Above: The theme for 2003 was 6in scale miniatures and here some of the Maskell Foden wagons were posed up in front of the house for photography. There were over thirty half-size exhibits, four of which were shown with their full-size counterpart.

Right: The unusual shape of the smokebox of this engine is because it was built with a superheater fitted there. This is Garrett 6nhp engine No. 29764, AC 9326 *Olive* seen in 2004.

ANNUAL RALLIES ONGOING • 113

Above: The Howard ploughing engine was back in 2006 for the 'Made in Bedfordshire' themed event. Now fully restored it makes a contrast with the view taken in 1998.

Opposite above: Highlight of the 2007 event was the first gathering together of all seven of the surviving engines built by C.J. Fowell & Co. Ltd of St. Ives, Huntingdonshire. Two of these, Nos 91-2, CE 7856-7, are locally based in Bedfordshire and are regulars at this show, but one of the others came from Cornwall, and No. 98, CE 7858, was brought over from its home in Ireland. Here the seven are posed for photographers with the most modern, No.108 of 1922 nearest on the left. There was also a marquee display on the history of the company.

Opposite below: Visiting in 2007, *Johnboy* is the first of at least four new build engines erected by Dawson Bros, Bicker, Lincolnshire. The engines are based on a Foster 3nhp tractor design and *Johnboy*, completed in 2007 carries makers No. 14739 continuing the sequence of Foster numbers when they ceased production. As it is a new build engine it cannot carry a period registration and is registered GN07 UZE.

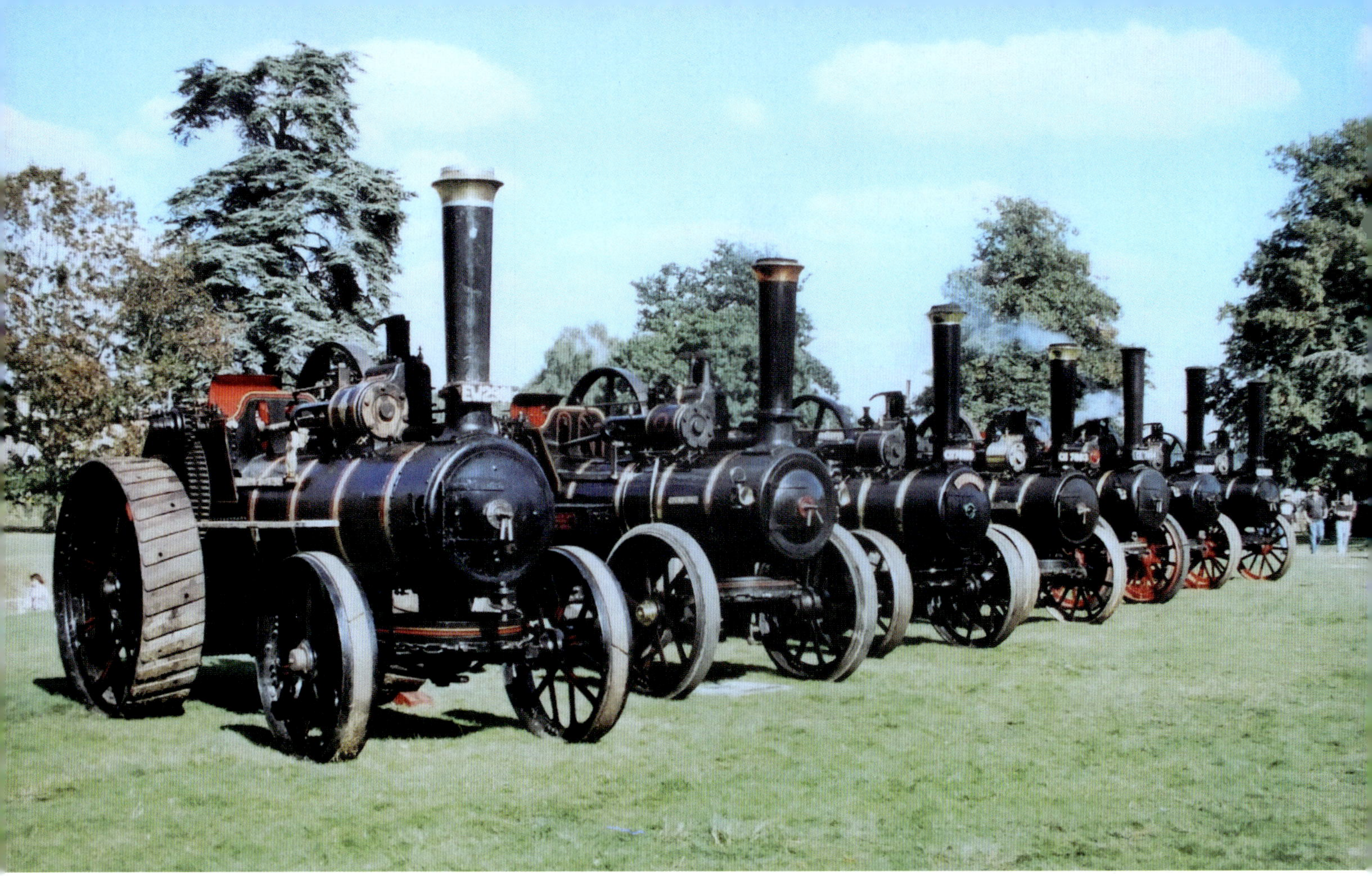

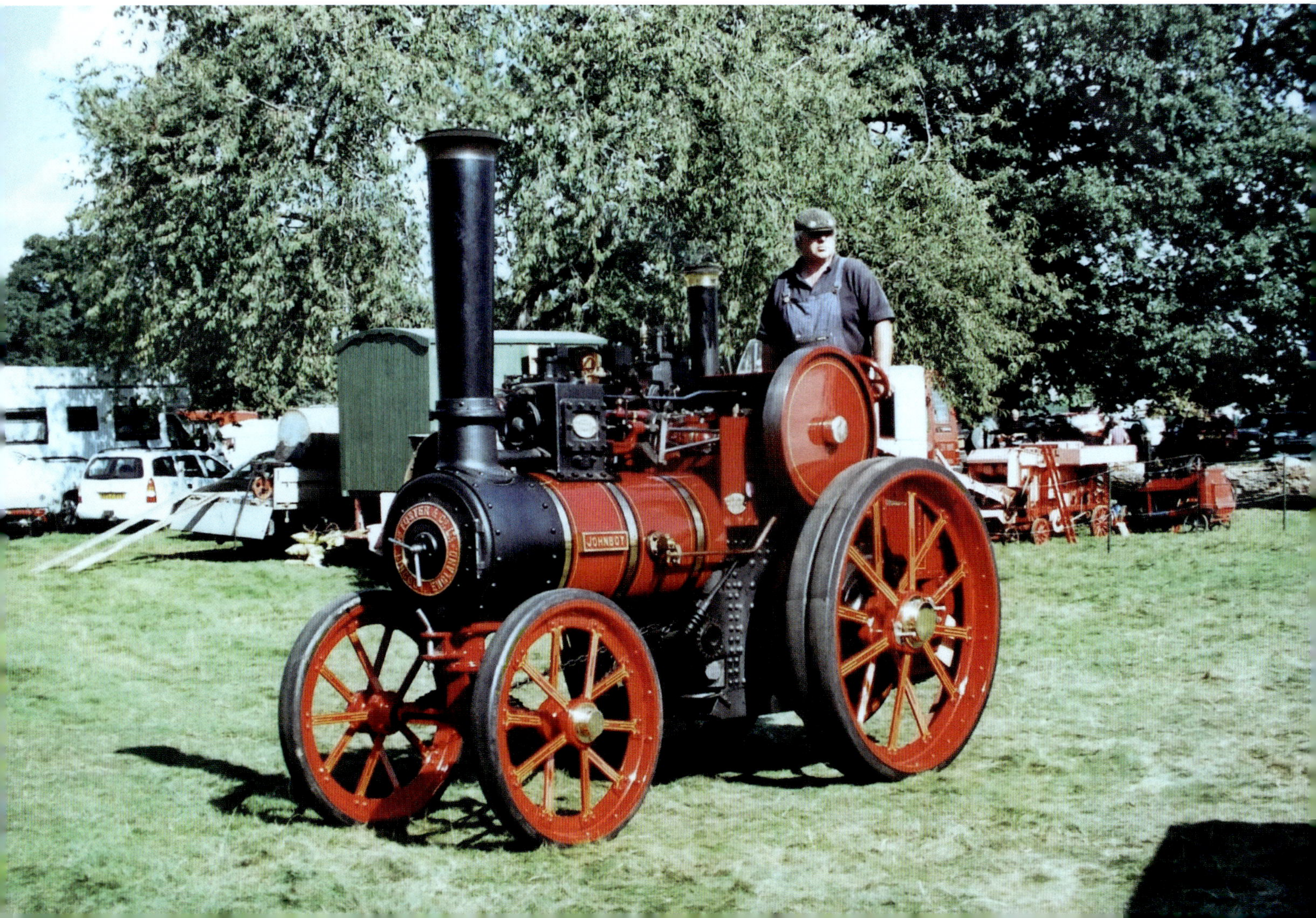
JOHNBOY

Above: Two engines that worked in Scotland are posed together at the 2008 show. RG 1417 is a 1930 Sentinel DG6 tipping wagon which worked on Aberdeen harbour. SO 2182 *Wizard* is Aveling & Porter No. 11137 of 1925 and worked for Moray County Highways Department.

Opposite above: The Aveling returned from Scotland for another visit the following year, now fitted with front registration plate and a canopy. This is a rare example of the KT type compound engine.

Opposite below: The theme for 2009 was crane engines, following the restoration of Burrell No. 3166 *Joe Chamberlain*. The engines were lined up in various combinations for photography and on the left is *Joe Chamberlain*, the paintwork not yet fully completed. Alongside is Burrell No.4074 *The Lark*.

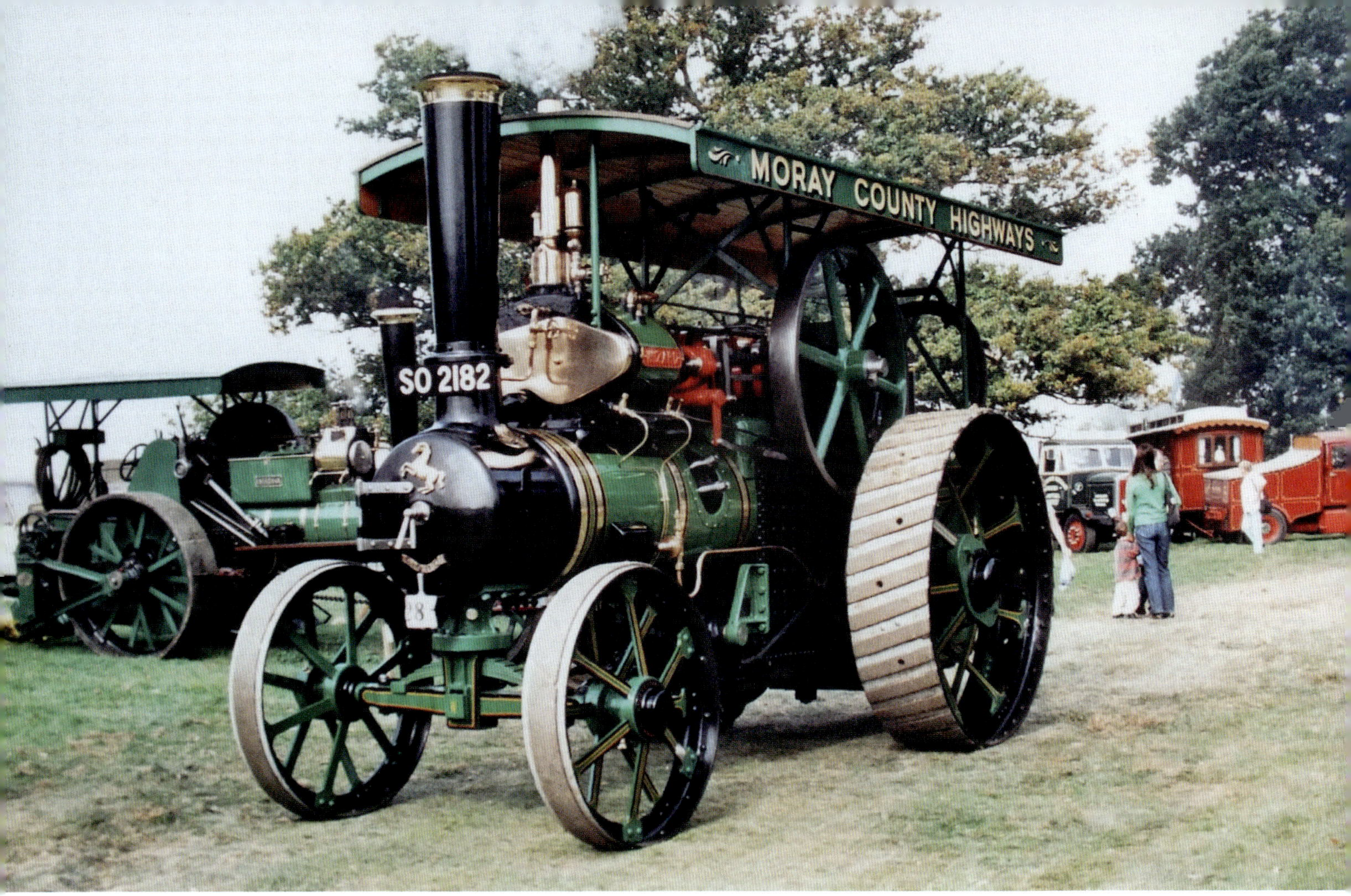
MORAY COUNTY HIGHWAYS
SO 2182

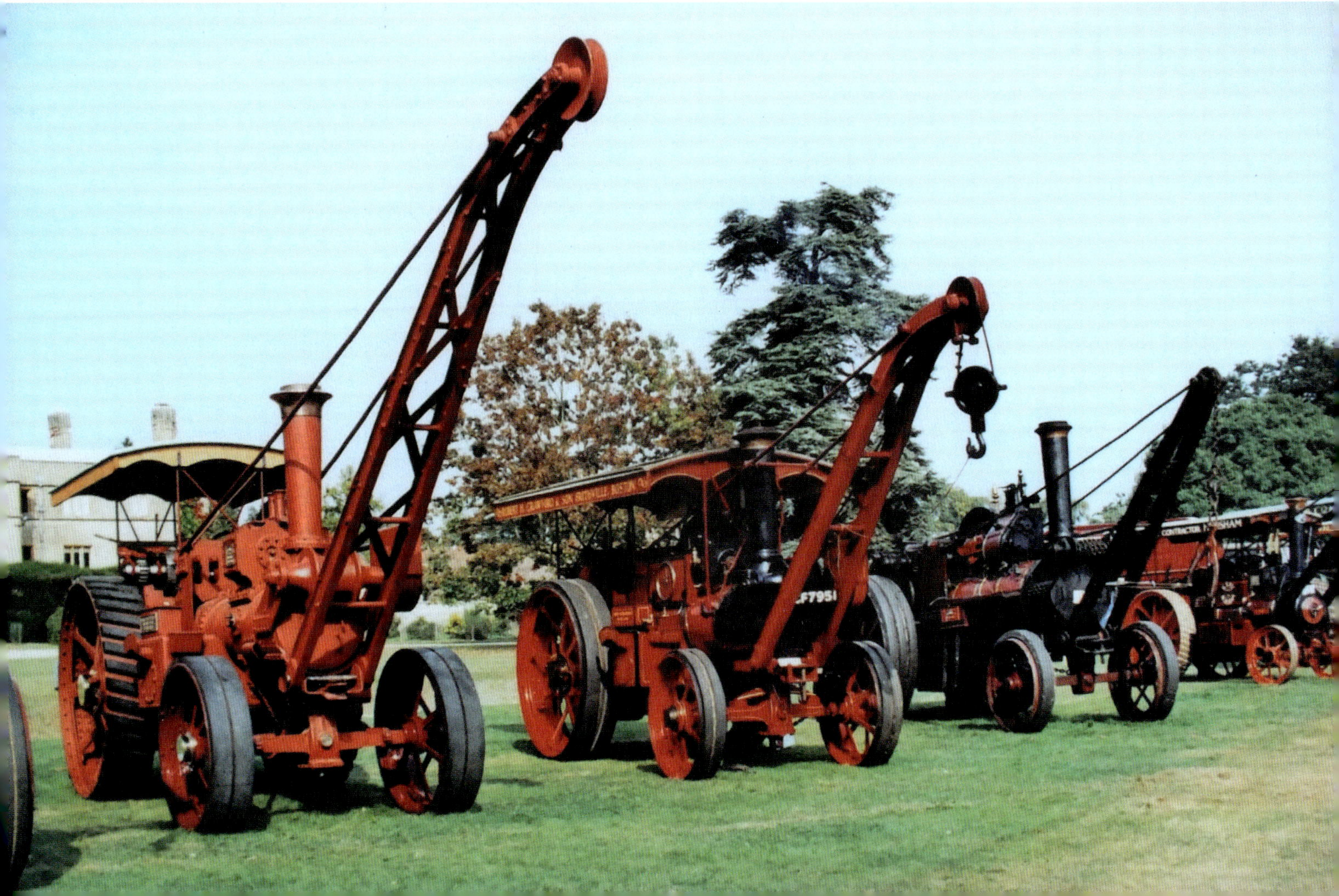

Above: Some tractors were made as convertibles with a bolted on front plate that could be removed and replaced by a mounting for a roller and the front wheels then replaced by a front roll. This was intended for use on large country estates where there was a requirement for both types of engine, but without the expense of buying one of each type. With the crane engines on hand, the opportunity was taken to demonstrate how the switch could be made using crane engines (although in practice, block & tackle would be more likely used). Aveling & Porter BP 7501 *Queen of Herts* has been propped up while the roller plate is attached, held in place by Fowler RF 6092 *Wolverhampton Wanderer*.

Opposite above: In the next stage, the front roll has been brought in by Burrell AB 8904 *Old Tim* and is being positioned in place before the tractor is lowered down by *Wolverhampton Wanderer* for the holding pin to engage.

Opposite below: Showman's engines were traditionally painted maroon unless the customer specified otherwise. In 2009, we see the unusual sight of newly restored Burrell No. 3910 *Wait & See* painted in yellow as delivered in 1921 to Crowther & Johnson of Leeds to work with their 'Flying Pigs' ride. However, in 1923 it passed to Pat Collins as displayed on this side of the canopy.

SCREEN BROS. LTD. OLDBURY. BIRMINGHAM.

PAT COLLINS. TOURING AMUSEMENTS.
DH 3544

Above: In 2011, the theme was the fiftieth anniversary of the making of the film *The Iron Maiden* in 1962. BSEPS arranged a special commemoration as their long-standing President, John Crawley, who had recently died, had been the technological advisor for the film and had arranged the supply of the engines. The star of the film, Fowler showman's road locomotive No. 15657 *Kitchener/The Iron Maiden* was duly present, visiting from its home in Scarborough. This was posed with a replica of the barn which the runaway engine crashed through in the film, with John Crawley as 'stunt double' for actress Anne Helm. Two other showmen's road locomotives that featured in the film were also present along with other engines that also had a screen appearance.

Opposite above: The theme in 2012 was wagons and a wide variety was present, representing the various approach to their design by different makers, some of whom soon lost out to their more successful competitors. One such was E.S. Hindley & Son of Bourton, Dorset. FX 33 is a replica of one of their early wagons completed in 2010 and using some original parts. No protection for the driver – they had to be pretty hardy in those days!

Opposite below: A highlight of 2014 was the return to steam of one of the two surviving Allchin rollers, both of which served for their home town of Northampton. Allchin No. 1131, NH 3416 was built in 1900 for Northampton Borough Council as fleet No. 1 and was followed by No. 1187, NH 3417 in 1901. Both survived in Council ownership after withdrawal. In 2005 The Borough Council offered the unrestored NH 3416 on lease to preservationists on condition that the roller would be restored and maintained at the preservationist's expense and would be painted in Council livery and made available for certain civic events. The 10-ton Allchin was restored at the Northampton & Lamport Railway.

E.S.HINDLEY & SONS
MAKERS
BOURTON
FX-33
25

NH 3416
COUNTY BOROUGH OF NORTHAMPTON. HIGHWAYS DEPT No1
101

Above: In 2014, another of the Saunders collection, Foden No. 13008, displays its revised livery style – previously it was blue. This was built in 1928 as rigid 6-wheel wagon registered TU 9235. It was later converted to a tar sprayer and as such worked for the Limmer & Trinidad Lake Asphalt Company. It was converted to the present shorter tractor form in the 1960s. This came to the Saunders collection in 1987. Now registered SS 9191, it does not usually display this.

Opposite above: Old Warden is home to the Shuttleworth Collection. A resident part of the collection is Clayton & Shuttleworth No. 46817 *Dorothy* of 1914. Built as a convertible traction engine it was used in roller form for much of its working life with Messrs Buncombes of Highbridge.

Opposite below: Appearing in 2014 was McLaren No. 1295. This 8nhp engine, now registered BF 5927 and named *Mr Tweedie*, was repatriated from Argentina in 2002.

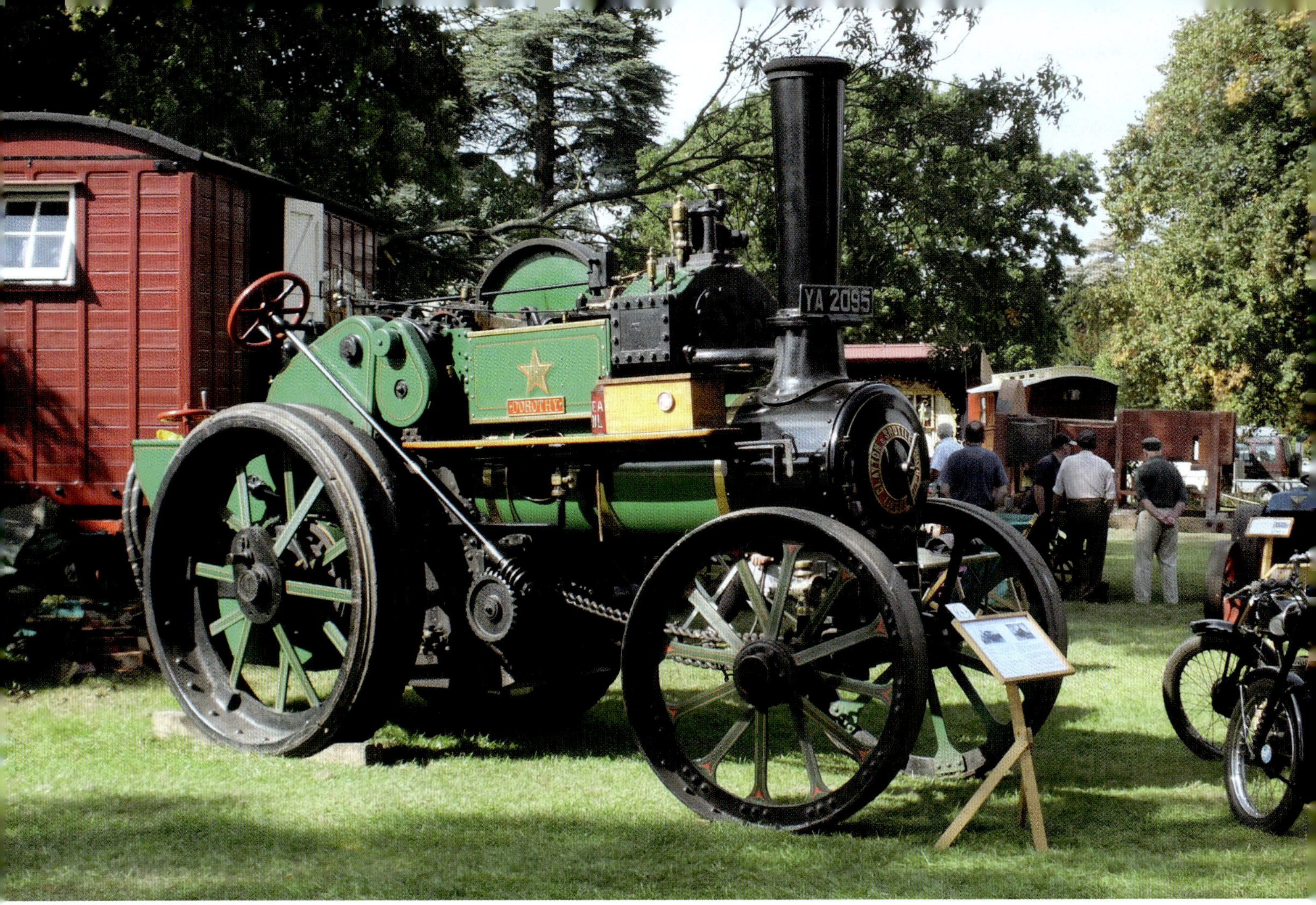
YA 2095

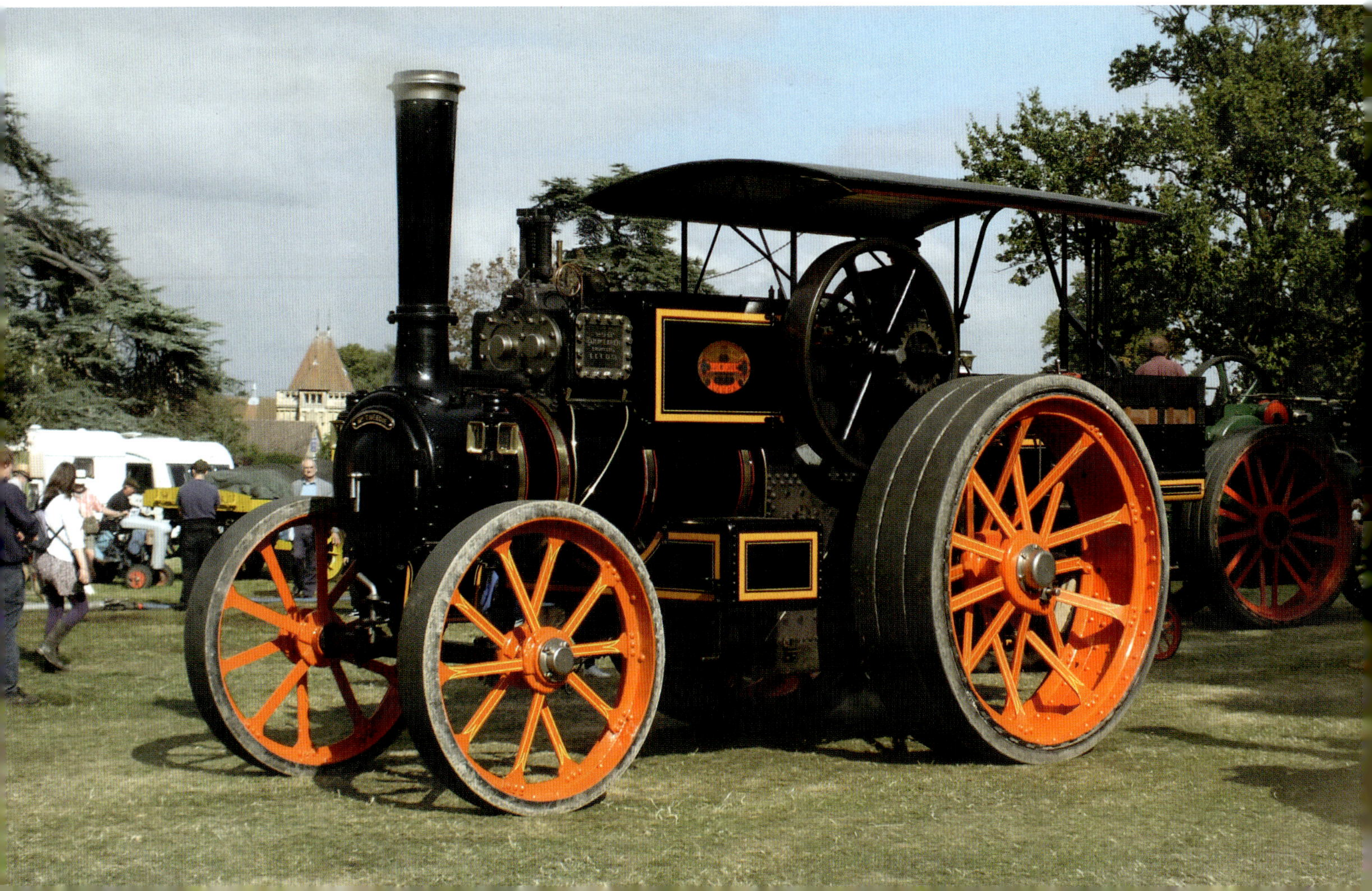

Above: Foden articulated tractor No. 13536 was entered from its home in Durham in 2016. This had been rebuilt to its original form in 2015 after many years as a tractor. Built in 1930, the original commercial owner was F.H. Richards of Leicester.

Opposite above: The ultimate in steam wagons. There are only two 8-wheel Sentinels listed, this DG8P and one S8 type (see p. 53). UX 5355 dates from 1929 and was rebuilt from surviving parts but requiring a new chassis. It is posed for photographs at the 2018 show.

Opposite below: They don't come much bigger than this! The massive Ransomes, Sims & Jefferies No. 22743 was built in 1910 as a Colonial type direct ploughing engine and was repatriated from South America via Preston Services. Contrast its appearance now restored in 2018 with it at Preston in 2011. (see p. 86)

TARMAC LIMITED.
ETTINGSHALL. WOLVERHAMPTON.
10
TELEPHONES
41101/8 BILSTON.
TARMAC LIMITED
ETTINGSHALL.
WOLVERHAMPTON.
TARMAC
UX-5355

D.PITTOCK Engineering LD MERSEA ISLAND

CHILTERN STEAM RALLY 1962–DATE (NTET AUTHORISED EVENT)

The Chiltern Traction Engine Club was founded in 1961 and reconstituted in 2003 as the Chiltern Traction Engine Club (2003) Ltd. The first rally was held at Chartridge End Farm, Chesham on 30 June 1962. This site was used until 1965 with the 1966 rally being held at Tring Park. Other venues followed until the present site at Prestwood, near Great Missenden, Buckinghamshire was adopted in 1985, the rally now being held on the first weekend of July.

Opposite above: A rally held at Gerrards Cross on 10 September 1972 saw Aveling & Porter tractor No. 11486, KM 7100 *Morning Star* amongst the entrants.

Opposite below: A most unusual beast was VN 2092 *Hunslet*. Aveling & Porter No. 2068 is described in the Traction Engine Register as a 4nhp tractor rebuilt from a roller in 1970 and would later be rebuilt again in 2001 when it received conventional traction engine wheels. At the time, it was lettered as belonging to Robert Oliver Ltd, Bishop Auckland, Durham so what was it doing in Gerrards Cross? It was listed as located at Chester-le-Street in 2012 but had relocated to Scotland by 2016.

2
KM710

ROBT OLIVER LTD
BISHOP AUCKLAND
SEMENTS
FOR HIRE
APPLY
HUNSLET
VN2092

Above: Moving forward to 2003, this is 1920 Foster 5-ton wagon No. 14550 carrying names *Sir William* and *Tritton*. This was brought back from Australia in 1983 by Tom Varley. Foster & Co only built sixty steam wagons and this is the only survivor.

Opposite above: In 2005 we see Aveling & Porter roller SX 923 *Jupiter* driving a stone crusher. This is one of the larger roller designs rated at 15 tons. It was originally preserved in 1965, attended three rallies in 1967-8, and was then laid up for twenty-two years before changing hands in 1991 and being restored over the next ten years.

Opposite below: Also in 2005 is rare Mann wagon U 4990 of 1919. The Mann's Patent Steam Cart & Wagon Co. Ltd engines were a product of Leeds, and this wagon like the Yorkshire wagon (see p. 26), also made in Leeds carries a Leeds registration number.

SX923

2
U4990
BEYER, PEACOCK & Co. Ltd.
Gorton Foundry, MANCHESTER

Above: Very much a local engine, NM 257 *Britannia* hails from Chesham and has been a veteran attender of both this and the Knowl Hill rallies. Burrell No. 2668 was built in 1904. Here it stands by the fairground carousel ride. Also just visible is the rear end of a Scamell 'Showtrac' lorry – a design built as a direct replacement for the showman's engine.

Opposite above: This part of the Chilterns is noted for timber and chair making so there is always a good selection of displays relating to this. Working one of the sawbenches in 2010 is Marshall No. 61970 *Emma*.

Opposite below: A fine selection of timber handling machinery is usually on show, both steam and internal combustion. In the ring in 2014 we see Foden products of both types. AMB 300 is a timber tractor from 1933, the most recent steam engine by this make recorded in the Register. This had been repainted in its original owner's livery. Alongside is GV 8092, a 1941 Foden STG5 timber tractor which is fitted with a Gardner 5LW engine.

BJ5934
25

JAS. MURCH & SONS.
CONTRACTORS.
UMBERLEIGH.
DEVON
AMB 300
Foden
T. REYNOLDS
Timber Merchant
BURY. ST EDMUNDS
Foden
Diesel
GV 8092
99

Above: A surprise entrant in 2014 was the Bicknell family's latest restoration – Fowler Road Locomotive no. 12257. This was new in 1910 to the Timbrebongie Shire Council in Australia and was repatriated in 1988. This replaced the well-known Ransomes Sims & Jefferies crane engine *Hooky* which was listed in the programme to attend. The Fowler was back in 2015 by which time a canopy had been added.

Opposite above: Activity in the sawing area in 2015 and this time power is supplied by Aveling & Porter AF 6001 *Jubilee*, a local engine from High Wycombe. This engine had been owned commercially by the famous contractors R. Dingle & Sons Ltd of Stoke Climsland, Cornwall.

Opposite below: Another local stalwart is 1897 Fowler traction engine No. 7788 *Black Jack*, registration BH 6971. This was bought new by the Boughton family and worked commercially by them until the 1950s before entering preservation. It returned to the rally scene in 2017 after an absence for overhaul.

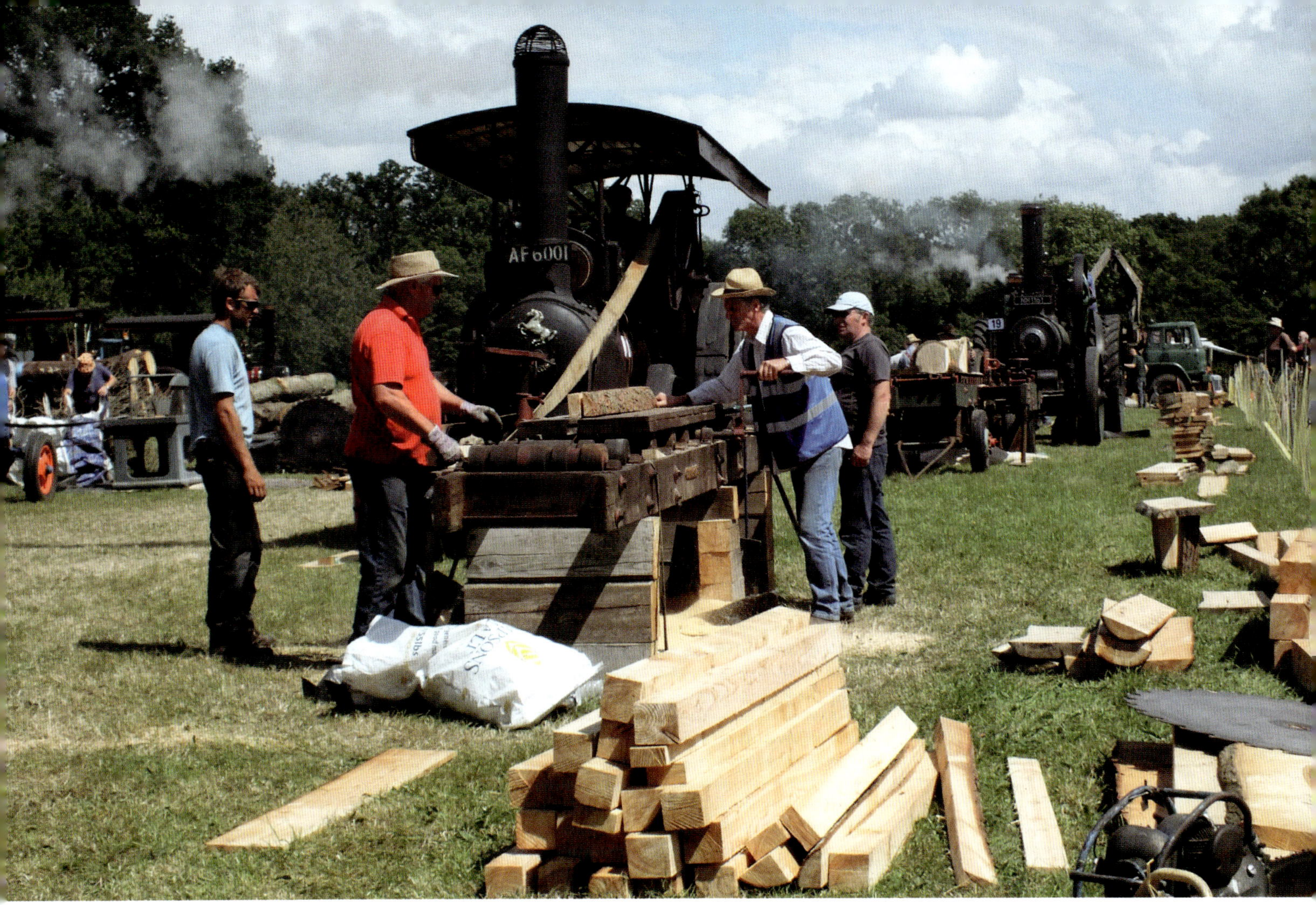
AF 6001

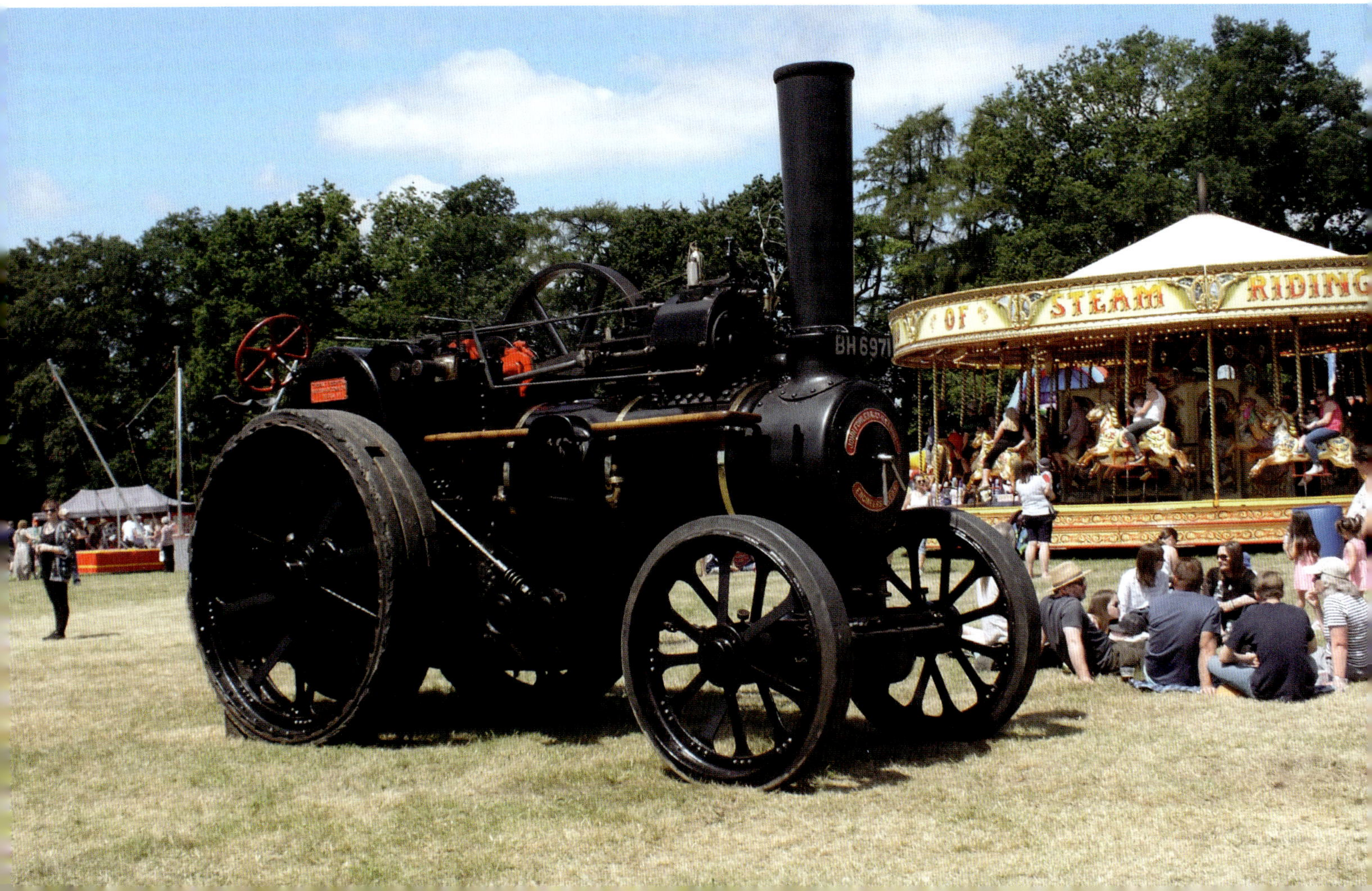
OF STEAM RIDING

HERTFORDSHIRE RALLY 1964–DATE
(NTET AUTHORISED EVENT)

The Hertfordshire Steam Engine Preservation Society was formed on 10 July 1964. Although there had been some locally promoted rallies in Hertfordshire before this, the first rally sponsored by the Society was held at Six Tunnels Farm, Gaddesden Row in September 1964, when twenty-three full-size engines were present. The rallies were held at various sites in subsequent years including Great Wymondley, Knebworth Park, Pirton and Little Wymondley. In the 1970s, several rallies were run in the summer months each year. An annual rally is now run by Herts Steam Club Ltd on behalf of the Society.

From at least 2003-07 it was held at a site on Leighton Buzzard by-pass (2008 was cancelled due to poor ground conditions). In 2009, the rally moved to the Oaklands College Smallford Campus at St. Albans where it has remained subsequently. The rally is now known as the St. Albans Steam & Country Show.

Below: On 15 July 1979, I visited a one-off rally held in the grounds of Hatfield House. I don't know for sure if this was organised by the HSEPS but it seems likely. Towards the end of the day, Marshall No. 43560 *Winifred* of 1905 has been placed onto Fensom's AEC Matador low loader for the journey home. The Fensom family of Colmworth, Bedfordshire purchased *Winifred* for commercial work in 1946 and later preserved it, attending all the early Bedfordshire rallies. Although sold to new owners in 2012, the engine remains a regular rally attendee.

Above: At Little Wymondley in 1989 is Aveling & Porter roller No. 7385 of 1911 towing a tar boiler. This little roller, weighing only four tons was built for the City of Oxford to work in some of the narrow lanes of the university city.

Below: Following it was 1902 Marshall No. 36258, FU 92 *Punch*. This was converted from a roller in 1972. Hailing from Biggleswade, it is a regular at both the Hertfordshire and Bedfordshire rallies, often driving the threshing machine here being towed.

Above: At St. Albans in 2011 is a regular of the Hertfordshire and Bedfordshire rally scene, Burrell No. 4055 *Crimson Lady*. This is a double crank compound engine dating from 1927. This was displayed on Burrell's stand at the Royal Norfolk Show when new. It has been with the Fensom family since 1959.

Opposite above: Highlight of the 2013 show, and No. 1 on the programme was former St. Albans Corporation steam roller NK 1102 which was celebrating its centenary. This was the third roller purchased by the Corporation and remained in service until 1958 when it was sold for £40 to a local enthusiast. Now owned by Chris Griffiths of Tring, this is Aveling & Porter 'R10' type 10-ton roller No. 8064.

Opposite below: Appearing in 2015, making a first rally attendance for many years, was Sentinel DG4 wagon No. 8070 of 1929. TE 9662, now owned by John Forshaw of Clifton, was supplied new to a coal merchant in Oldham in 1929. This was originally a three-way tipper but was later converted to a tar-spreading vehicle.

A.S. CALDOW & Co ROLLING CONTRACTORS ENFIELD MIDDLESE
NK 1102
St ALBANS CORPORATION

Sentinel
DG4
TE 9662

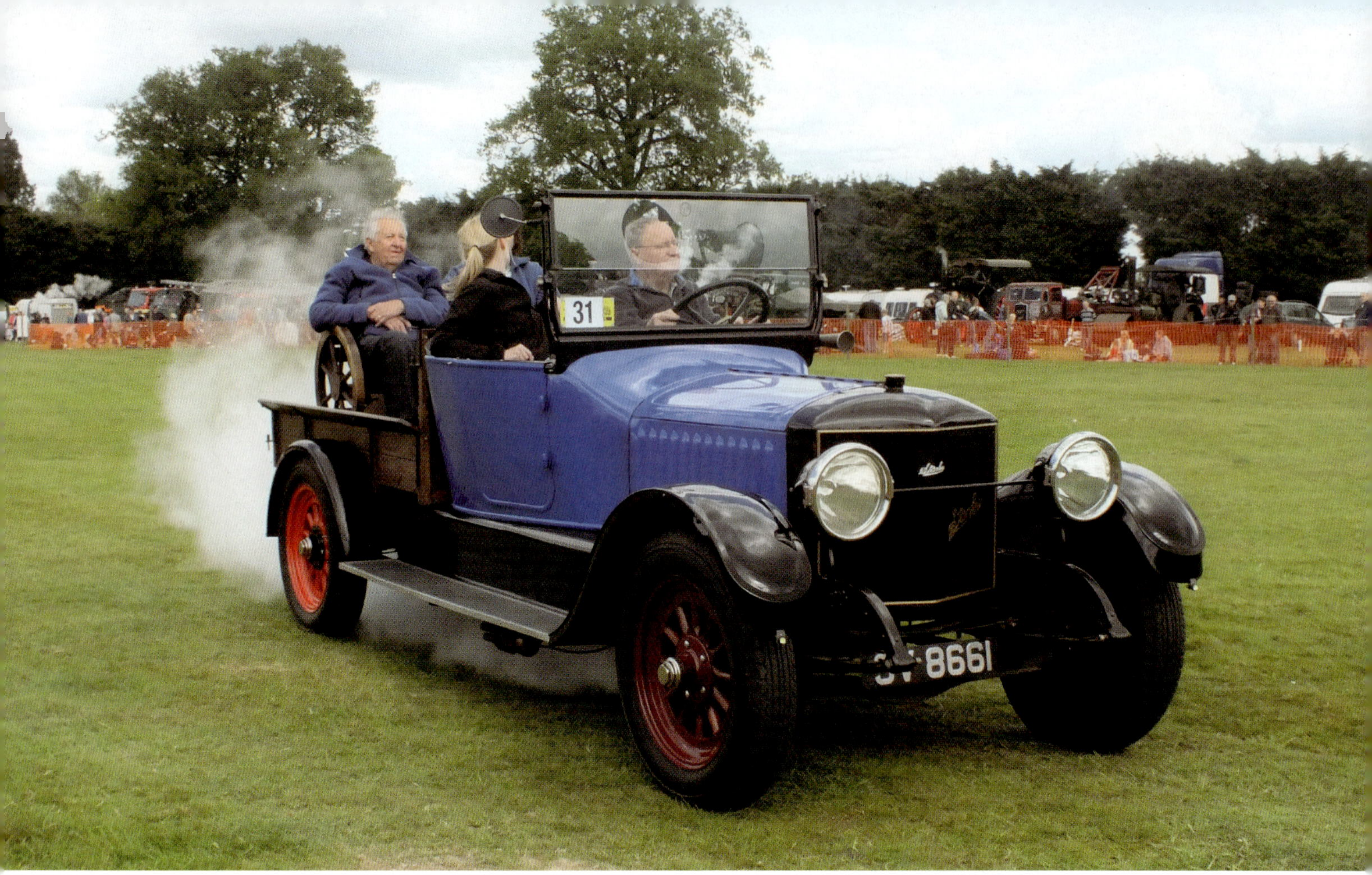
31
8661

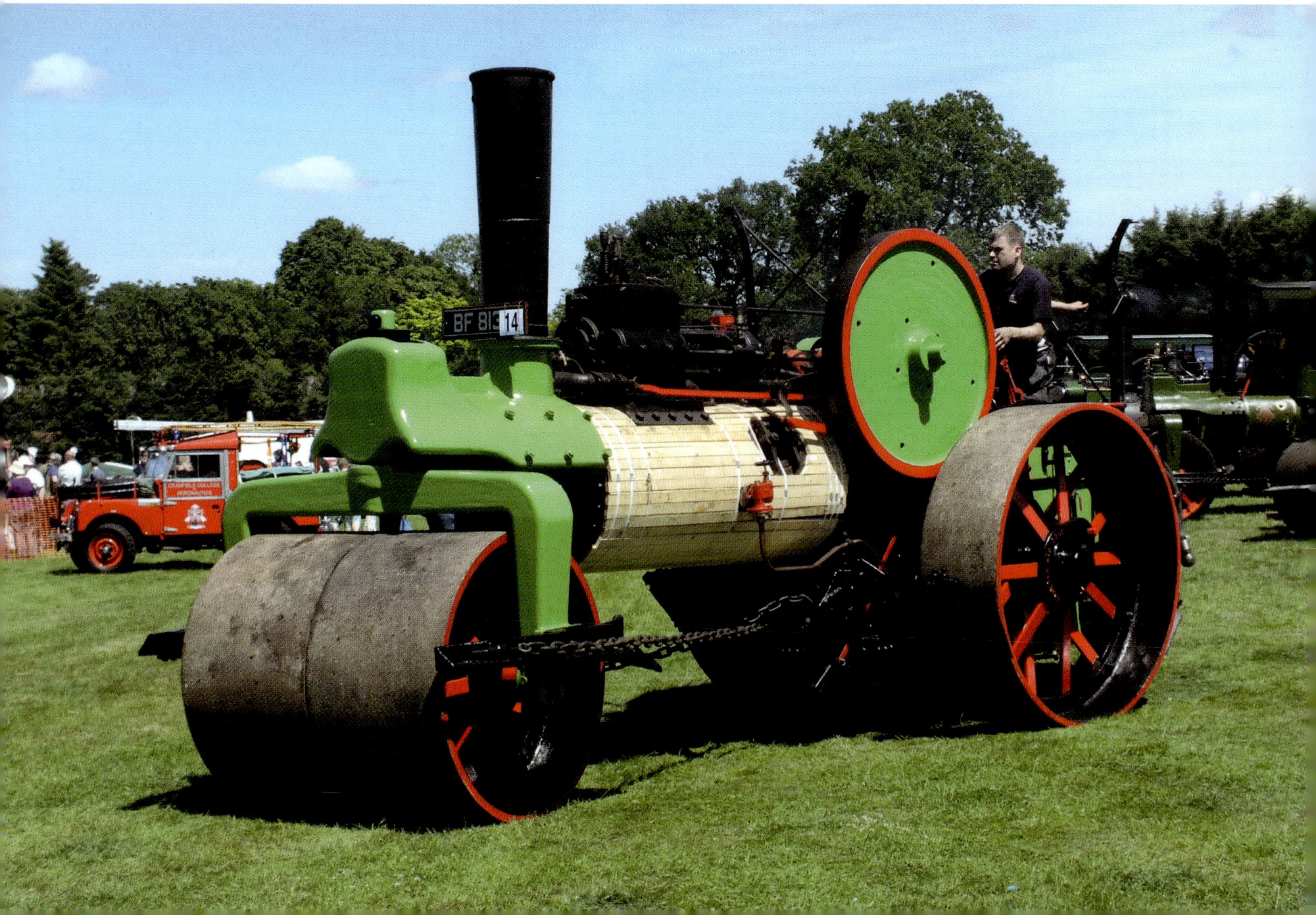
14

Opposite above: Making quite a stir when it was demonstrated in the ring was this 1916-built Stanley type 725 steam car now owned by John Savage of Wheathampstead. Originally built as a 5-seat car, its current pick-up body prompted the rally commentator to describe it as 'Britain's only steam powered SUV!'

Opposite below: Making a first appearance in 2017 was German built Zettelmeyer roller No. 302 built in 1926. This had been imported to the UK in bits and was in the course of rebuilding. Although in steam, it lacked the boiler cladding at this time – this had been applied by the time it was back in 2018. Owners are the Perrins family of Dunstable.

ISLAND STEAM SHOW 1975–DATE

The Island Steam Show has traditionally been held over the August Bank Holiday weekend at Havenstreet Station on the Isle of Wight Steam Railway which opened as a heritage railway in 1971. Until 2016, this was the weekend before the Great Dorset Steam Fair and some engines would often stop off at the Isle of Wight before going on to Dorset, thus adding to the local variety.

Below: The Isle of Wight has a number of resident engines and one that is actually kept at Havenstreet is 1905 Allchin No. 1415, AP 9077 *The Havenstreet Queen*. This was photographed in 1985.

Above: Visiting in 1985 was Aveling & Porter No. 12023. This unusual machine is a tandem roller with a vertical boiler and twin cylinders dating from 1928. The Irish registration number is due to this having been originally supplied to Ballyforden Quarries Co., Armagh. This engine was exported to Europe in 1987 but has since returned and is now kept at the Amberley Museum in Sussex.

Opposite above: The registration code for the Isle of Wight was DL, and here we see a genuine island registered engine. DL 2108 *Faithfull* is Marshall No. 52121, a 6nhp engine of 1909. This was taken at Havenstreet in 1986.

Opposite below: A visitor in 1987 was Burrell showman's road locomotive No. 3926 on its way to the Great Dorset. At the time this was Dutch-owned and carried the name *Stokomolief*. It subsequently returned to England and now carries the original name *Margaret* and registration NO 4999. It was built in 1922 for Henry Thurston.

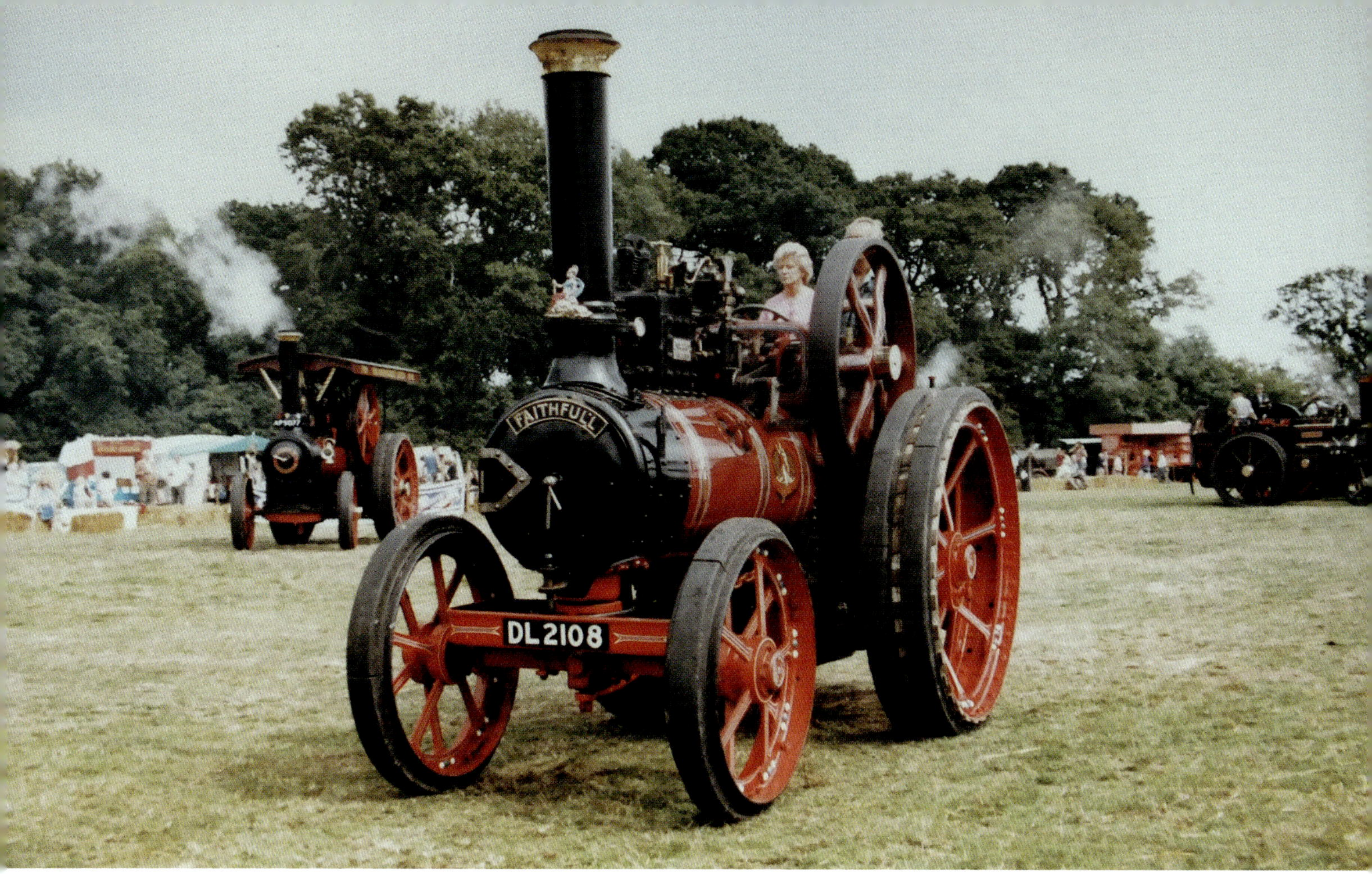
FAITHFULL
DL 2108

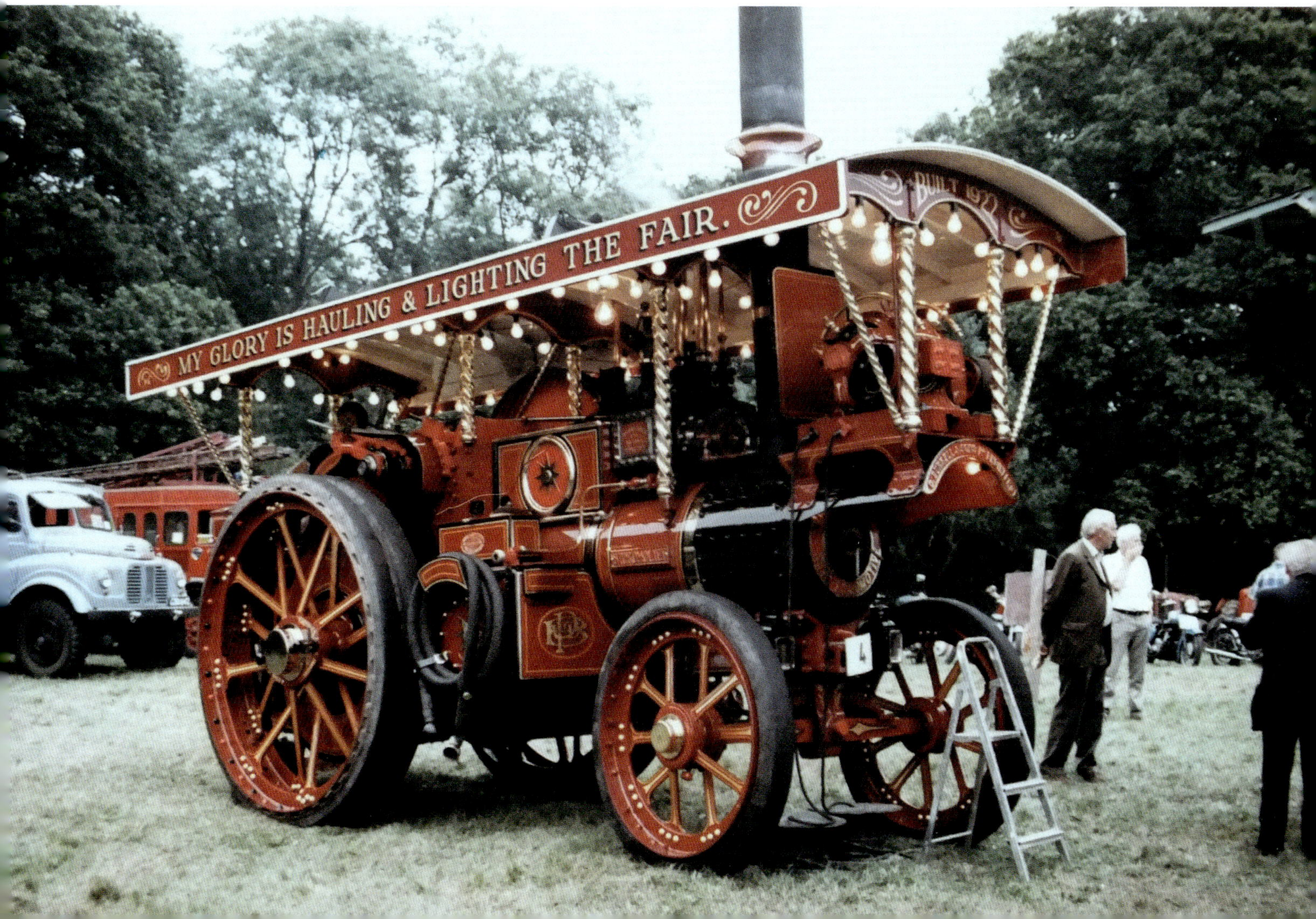
MY GLORY IS HAULING & LIGHTING THE FAIR.

Above: Between 1980 and 1985 the Foreman family of Romsey built a full-size replica of a Wallis & Steevens 3-ton tractor. Named *Our Baby* and given the registration number D716 JRU this is how it looked at the time, visiting the Island in 1987. The building of this engine was the catalyst for the spate of 'new-builds' that has followed since then.

Opposite above: Resident at the time, but since moved to Cornwall is Aveling & Porter 6nhp engine No. 4157, MA 5525 *Elizabeth* dating from 1898. This was taken at the 1989 show.

Opposite below: From 1995, we see Babcock & Wilcox roller YB 5089 *Brutus*, built in 1926. Rollers badged as Babcock & Wilcox were actually made by Clayton & Shuttleworth Ltd. who were absorbed by Babcock & Wilcox in 1924. This is another island-based engine.

MA 5525
ELIZABETH

BABCOCK & WILCOX ROLLER
THIRD PRIZE
7
Rank Xerox
The Document Company

Left: Powering the fairground ride in 2005 was Fowler showman's road loco UB 9763 *The Lion*, visiting from its home in the Salisbury area before going on to the Great Dorset Steam Fair.

Below: Since the Great Dorset Steam Fair changed its dates to clash with the Isle of Wight event over the August Bank Holiday, the island show has lost out on visiting engines and it is mostly resident engines that feature. In 2016 we see OT 5327, a 1925 Wallis & Steevens 'Advance' roller. These were specially designed for rolling asphalt. They have two high pressure cylinders and are capable of instant reversing which prevents the engine sinking and forming a hollow in the road when reversing. Built in three sizes of 6, 8 and 10 tons; of the 272 Advance rollers built, there are seventy-seven surviving in preservation. The *Pride of Bangbourne* alongside is an 1899 Wallis & Steevens No. 2459, registered AL 9335 and has been IOW based since 1999.

MEDWAY FESTIVAL OF STEAM & TRANSPORT, CHATHAM HISTORIC DOCKYARD 2002–DATE

Chatham Dockyard was decommissioned by the Navy in 1984 and eighty acres of the site was handed over to a charity, The Chatham Historic Dockyard Trust, and opened as a visitor attraction. Several steam engines and other vehicles owned by the Historic Dockyard Steam Society became based there for restoration to take place. Various transport events took place, but the present title for the rally was adopted in 2002. This event has not taken place since 2019 due to the Covid pandemic

Below: Seen at Chatham on 7 May 1990 is Marshall 59435 of 1912. The Traction Engine Register notes that this was exported to Ireland in 2013.

Above: There are only two traction engines made by the Wantage Engineering Co. in the Register. No. 1522 *Pioneer* of 1908 was based in Kent in 1990 when it appeared at Chatham, but like the Marshall above has since moved to Ireland.

Opposite above: The wording says it all! Foden wagon No. 12228 *Britannia* of 1926 was fitted with this box van body c.1967 for a round-the-world trip. It is owned by Mr List-Brain, proprietor of Preston Services in Kent.

Opposite below: 57159 Forward to the 2003 Festival and on the run around the site is this three-quarter scale version of a Mann wagon.

BRITANNIA
THE FIRST STEAM POWERED
VEHICLE ROUND THE WORLD
BRITANNIAS GRAND GLOBAL TOUR
WL 910

GEORGE
STAFFORD
STANLEY COMMON
GOLD MEDAL
BLACK PUDDINGS
AWARDS
GEORGE
STAFFORD
BUTCHER OF
DISTINCTION
No3
HAULAGE
G628 KAH

Above: Seen in 2004, *Magwitch* is a full-size near replica of an 1869 Todd steam carriage. It is uncertain whether the original was ever built, but if it was, it would not have been very successful as the plans showed that it lacked a differential and reverse gear.

Opposite above: Also on show in 2004 was 1924 Garrett roller No. 34594. This was previously based in Ireland, where it was registered IT 644, and has since returned there.

Opposite below: Among the miniatures on display in 2004 was a fine pair of early Fowler ploughing engines.

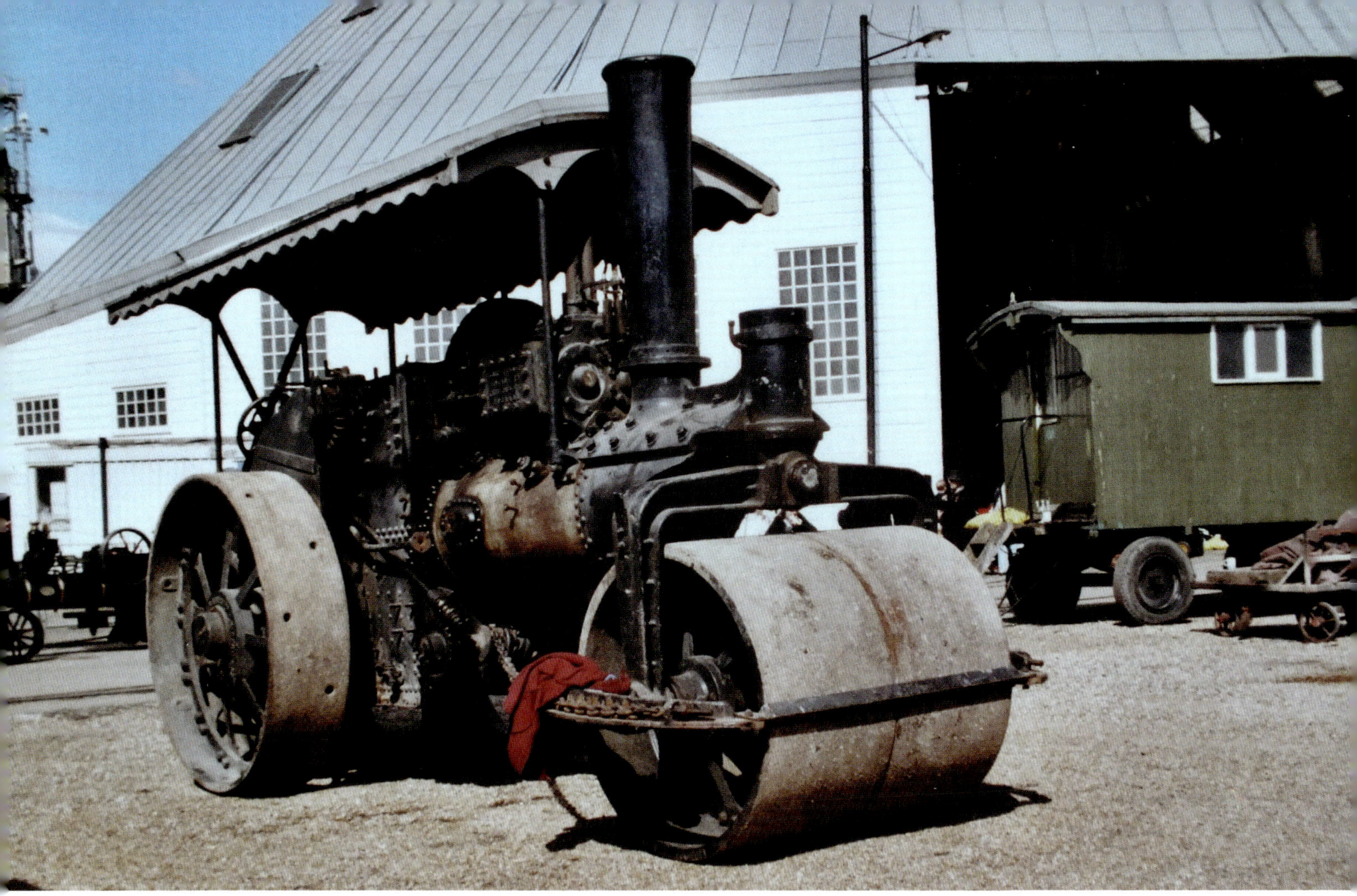

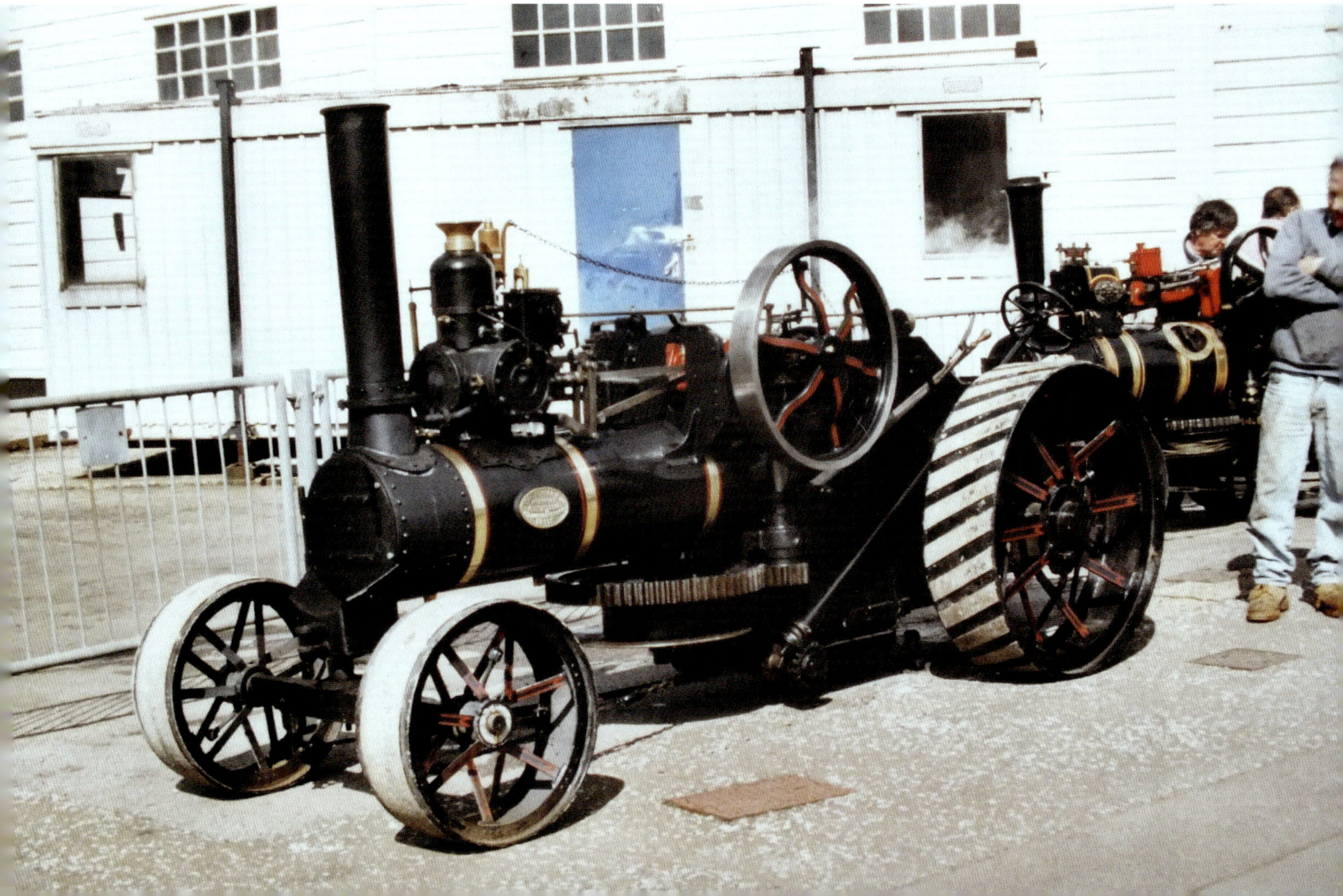

Above: Steam and beer always seem to go together, and so it seems appropriate to have two wagons in the liveries of breweries. Foden TW 4207 of 1926 promotes Newquay Steam Beers, while Sentinel S4 wagon 9208 of 1935, registered BYL 485 is in the colours of Shepherd Neame Ltd, Faversham – Britain's oldest brewery, established in 1698. Although by now promoting beer, TW 4207 started life with Edward B. Devenish of Rayleigh, Essex conveying a very different product – horse manure to market gardeners. It finished its working life as a tar sprayer.

Opposite above: Aveling & Porter traction engine No. 8632 of 1915 parades around the site on 25 April 2011. Now registered BF 4871 and named *Deborah Clair* this was reconstructed from a roller registered SV 601 and named *Queen O' Scots* in 2010.

Opposite below: The last time I visited Chatham was in 2013, when the number of engines attending was well down on the previous average. Amongst those present were Burrell No. 3739 *Pride of the West*, now resident in Kent and Fowler tractor No. 15732 *Sir Douglas* which was kept on site at Chatham. This was rebuilt from a roller in the 1960s.

BF 4871

AF 3836
THE BURRELL TRACTION ENGINE
JOHN FOWLER & Co (LEEDS) LTD
PT 832

NETLEY MARSH STEAM & CRAFT SHOW 1971–DATE (NTET AUTHORISED EVENT)

Another long-running event, organised by a local volunteer committee, and with profits aiding local charities and deserving causes. Netley Marsh is close to Totton and only a few miles west of Southampton. When it began in 1971 it was a one-day event called the Puff-in. The aim was to raise funds for repairs to St. Matthews Church. Following the success of this, it became an annual event, which now extends over three days.

Below: At my first visit in 1973 is Marshall roller No.71833, registration ER 697 and dating from 1919, which was with an owner from Dorset.

Opposite above: Also seen in 1973 is Burrell 7nhp agricultural engine No. 4053, TD 8047 *Dreadnought*. This was, and still is, owned by the Breamore Countryside Museum at Fordingbridge, Hants. Although both engines were entered from local counties, their original working lives had started out much further afield. ER 697 was registered in Cambridgeshire, while TD is a Lancashire registration. *Dreadnought* was supplied new to a Mr Thomas Robey of Melling, Lancs. in 1926.

Below: Taken in 1992, Garrett tractor BJ 3282 *Katrina* circles the ring against an impressive backdrop of engines that have already entered and lined up. A local engine, this is one of many Garrett tractors that have subsequently been given showman's fittings – in this case during the 1960s.

Above: Making an appearance at the rally in 2000 was Bolton's celebrity former steeplejack and TV presenter Fred Dibnah with his roller *Betsy*. This is Aveling & Porter No. 7632 of 1912.

Opposite above: Robey 4nhp tractor MF 3946 *Our Nipper* hails from Ringwood as stated on the canopy. This was taken at the 2006 Show.

Opposite below: Also seen in 2006 was this impressive miniature of the famous Hornsby Caterpillar track gun tractor. The chain track was patented in 1904 and Hornsby built one steam and three oil powered tractors which were demonstrated to the army, but with no orders forthcoming the patent was sold to the American company Holt who registered the 'Caterpillar' name as a trademark.

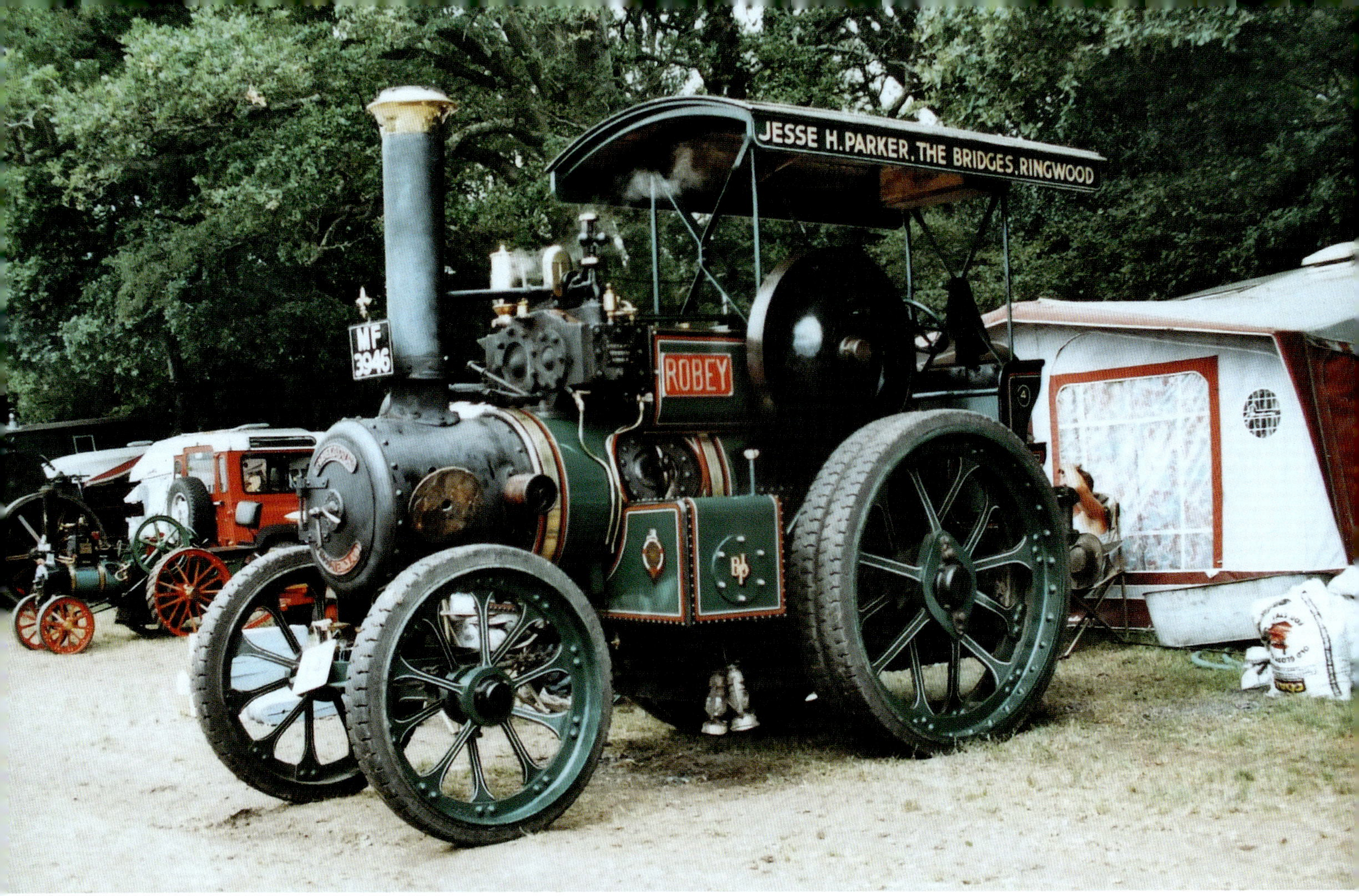
JESSE H. PARKER, THE BRIDGES, RINGWOOD
ROBEY
MF 3946

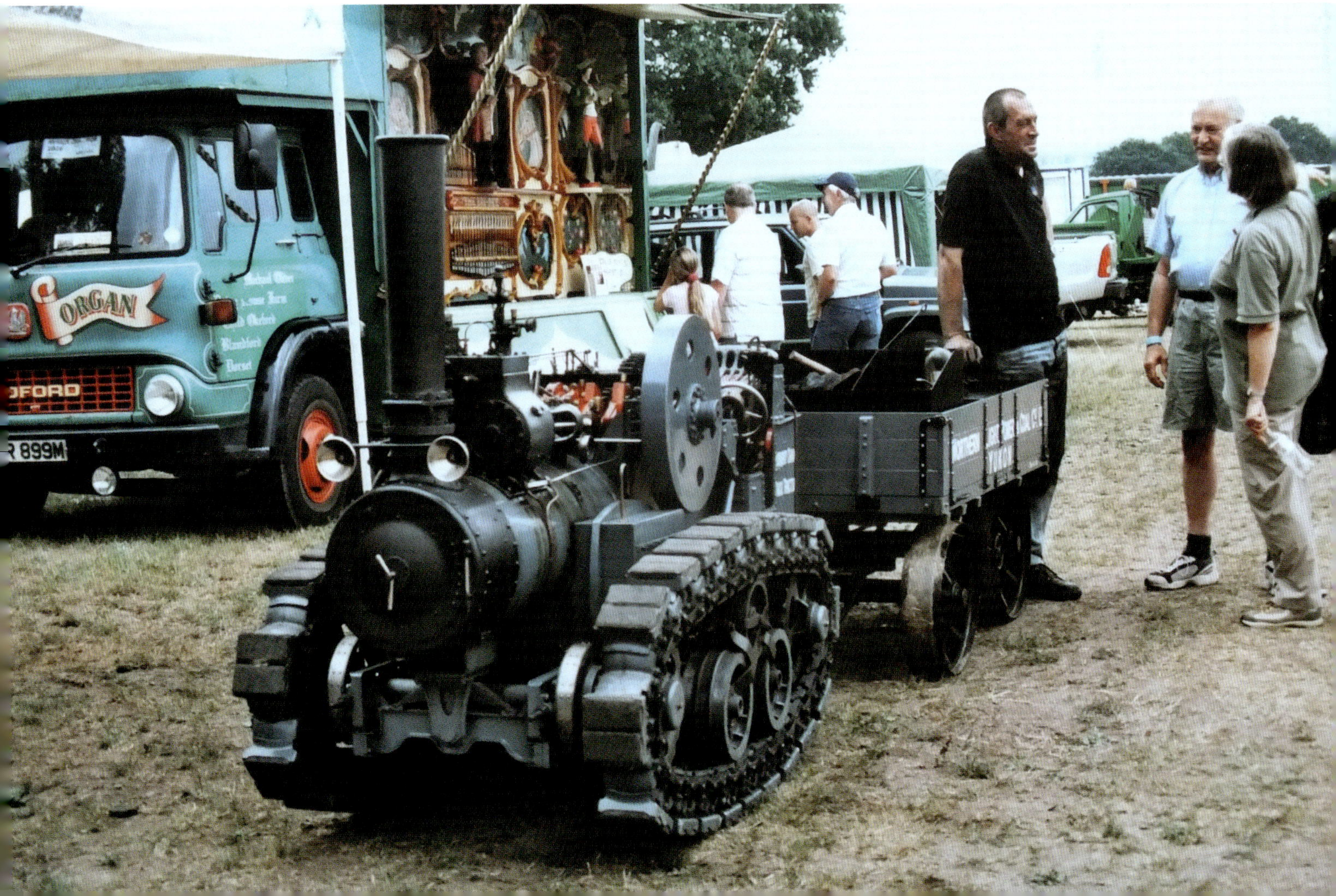
ORGAN

Above: Seen in 2014 sporting a recent repaint to the original owner's livery is 1929 Foden timber tractor No. 13444 *Little Lady*. From 1968 to 2014 this had carried showman's fittings.

Opposite above: Making a debut appearance in 2014 was Garrett showman's tractor No. 33505 of 1919. Built as a genuine showman's tractor, BJ 4308 *The Rambler* had been in Holland from 1978-2014. It was not in steam, so was being towed around the ring.

Opposite below: It's called '*Bitsa*' and that is what it is. This is a freelance showman's steam wagon owned by Robin & Rachel Neave of Alderholt. Originally built in the 1970s and acquired by the present owners in 2009, it has been rebuilt and now features a vertical boiler and a Bryan Donkin engine, as well as several Foden parts. Behind it comes Foden tractor KX 3340 *Samantha* of 1929 – probably one of the most prolific attendees at rallies. This was converted from a wagon during its working days.

THE RAMBLER
THE LOW MOUNTAINS STEAM COMPANY
BARRETT

Q705 LEL
OLD TIME STEAM FAIR
Bitza
3
KX 33 40
NEW FOREST CIDER
BURLEY

Above: Moving on to 2019 and another Foden wagon. M4257 *Tewlass*, No. 3210 of 1912 was a work in progress at this stage. This is recorded as being reconstructed from a tractor.

Opposite above: There is no questioning where this was built! Proudly sporting the Stars & Stripes is 1895 built Peerless engine *Princess America*. Peerless was the trading name of the Geiser Manufacturing Co., Waynesboro, Pennsylvania.

Opposite below: A regular entrant at Netley Marsh for some years has been *Titan*. This 6in scale Burrell was built in 1970 by a blacksmith in Shetland. During the day it was used to power a small saw bench and at night ran a dynamo for electricity in the home. It was sold in 1980 to an owner in Winchester and passed to the present owner in 1985.

1895
"PEERLESS"
8 NHP
PEERLESS
PATENTED

EST CIDER
RLEY
RPS 419Y
TITAN

WEETING STEAM ENGINE RALLY & COUNTRY SHOW 1968–DATE

The Weeting Rally is a long-standing event held on Richard Parrott's Fengate Farm, near the Norfolk/Suffolk border. Burrell engines are always a significant feature here, not surprisingly so close to Thetford where they were built. Richard Parrott owns several of the make, including 1877 built No. 748 *Century*, the oldest surviving Burrell in the UK. There is always a strong working element here with many engines belted up to sawbenches, threshing machines etc. Like the Bedfordshire rally, there has sometimes been a themed section such as a display of Burrell 'Gold Medal' tractors in 1997 to mark ninety years since the award of the RAC Gold Medal in the 1907 trials. 2018 saw the fiftieth anniversary rally with a number of engines that attended the first rally in 1968. There was also a nod back to the early days with some original programme features being repeated like the slow race and musical chairs.

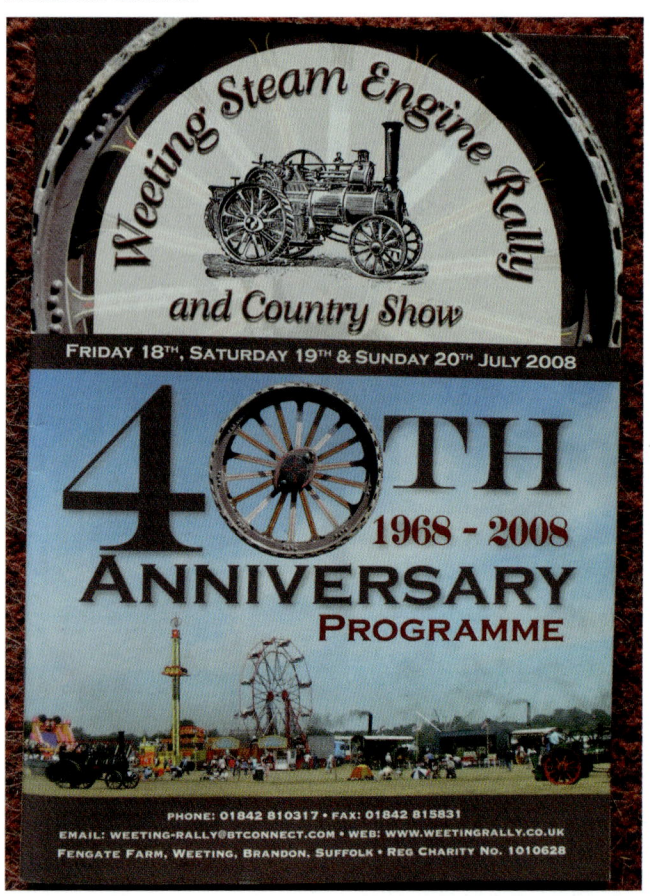

Opposite above: The oldest Burrell engine in preservation, CF 3667 *Century*, seen in 1997. This is a single cylinder 8nhp engine, weighing ten tons. It was originally built with chain drive but was rebuilt with gears prior to 1900.

Opposite below: Weeting is very much a working rally, and there are plenty of opportunities to photograph Burrell (and other) engines performing the tasks they were designed for. Resident Burrell No. 2948 *Dreadnought* of 1907 is driving a threshing machine in this view from 2003.

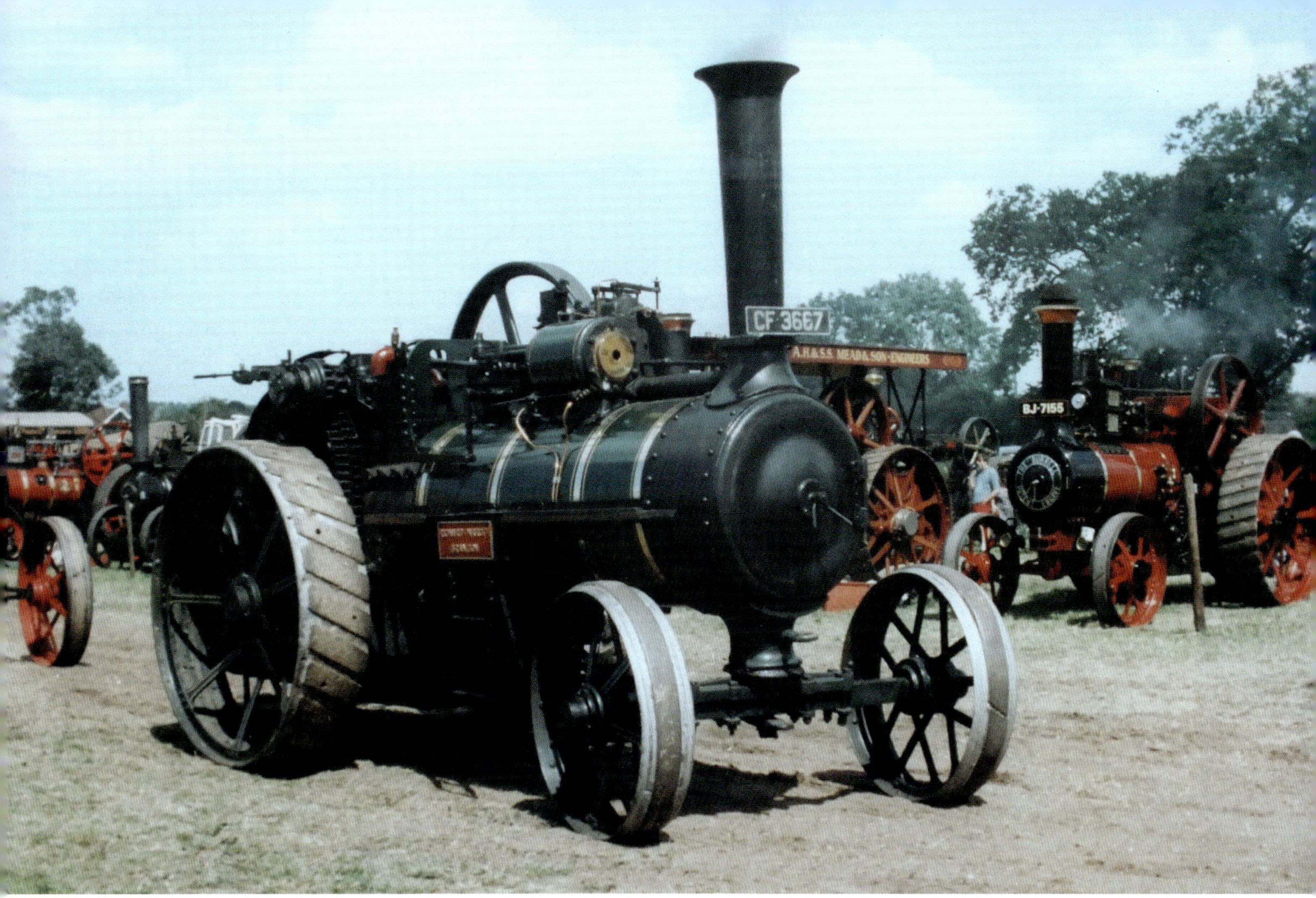
CF 3667
BJ-7155

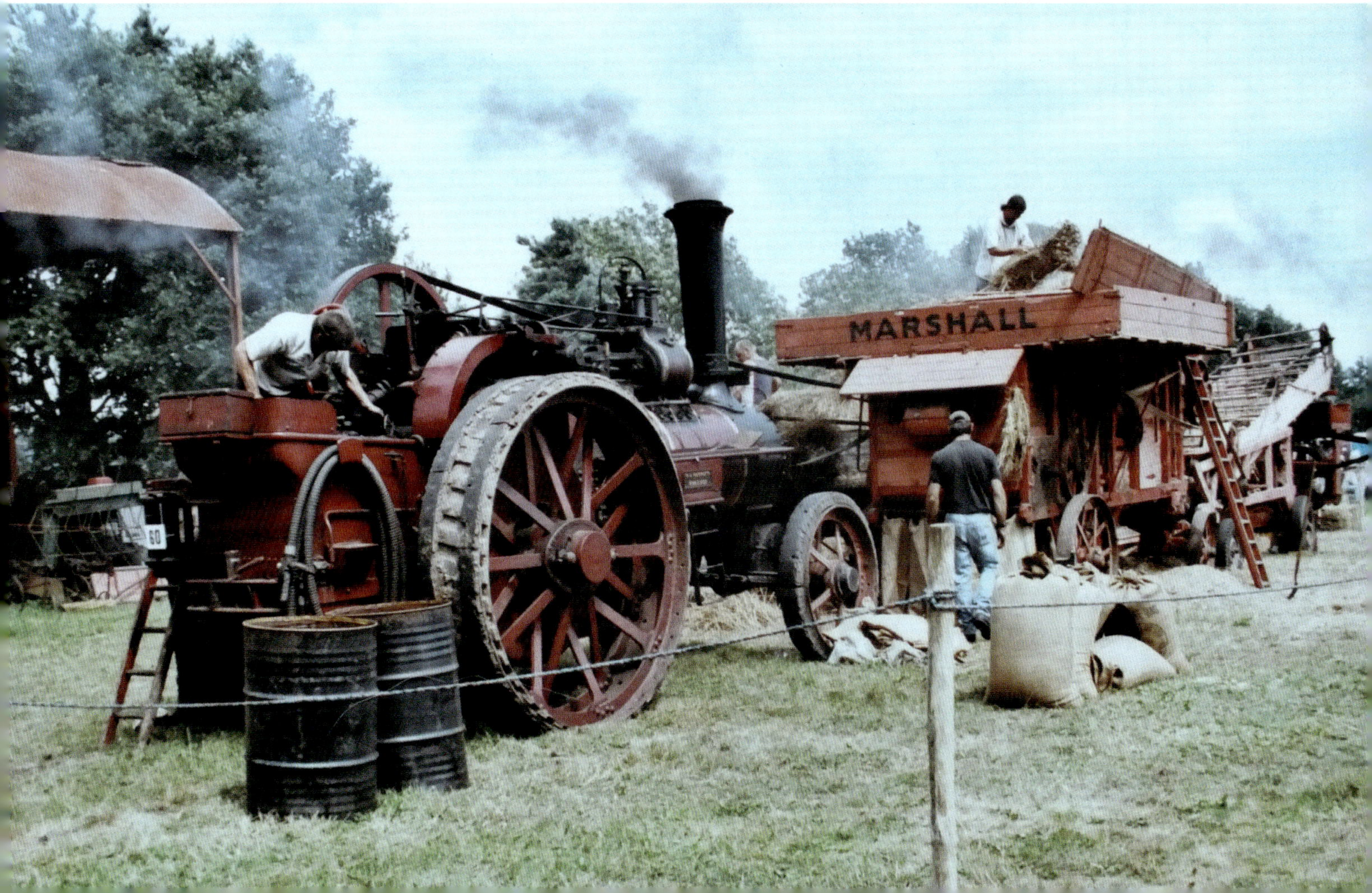
MARSHALL
60

Above: Weeting is also timber country with local forestry around Thetford. Resident Burrell crane engine TB 2740 *Lord Derby* positions a log ready for sawing in 2007. The engine was new to S.J. Rosbotham of Bickerstaffe, Lancashire for haulage duties. During the Second World War, it was used in Liverpool for clearing bombed property. The crane and jib were originally fitted to the Burrell works engine *Emperor*.

Opposite above: In 1993 a pair of sawyers take a break while demonstrating hand sawing a log in a saw pit. Behind is Burrell CF 7108 of 1926 belted to a sawbench. Incidentally, this is where the terms 'top dog' and 'underdog' come from. The sawyer above the log – the 'top dog' – was better off than the one below the log in the pit – the 'underdog' – because the latter would get all the sawdust falling down on him.

Opposite below: The 1993 rally brought together the four Burrell crane engines then operational. Here they are posed up in the ring. From left to right are No. 3197, AB 8904 *Old Tim*, No. 3829, PB 9687 *His Majesty*, No. 3695, TB 2740 *Lord Derby* and No. 4074, CF 7951 *The Lark*. No. 4074 was the last crane engine built by Burrells.

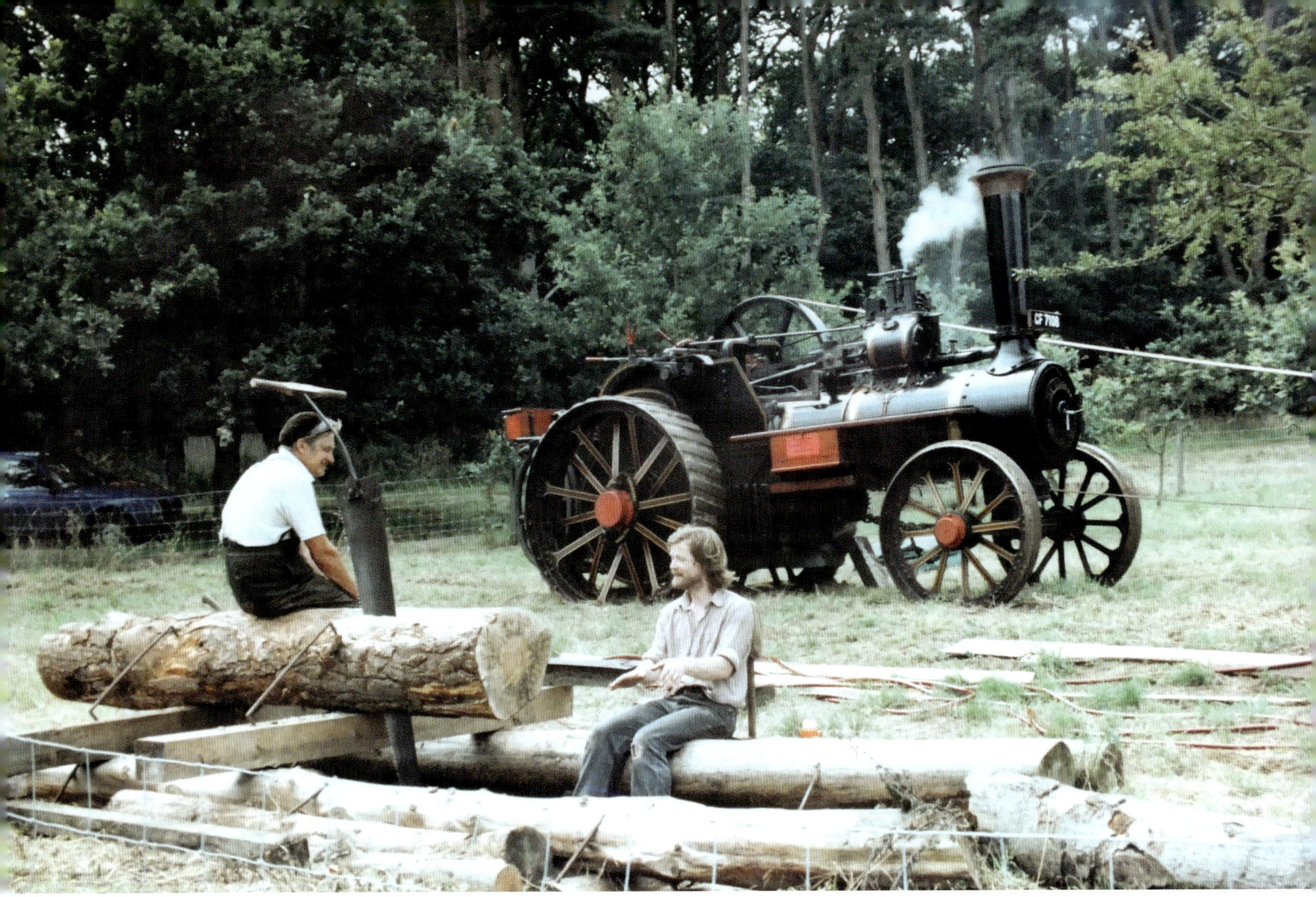

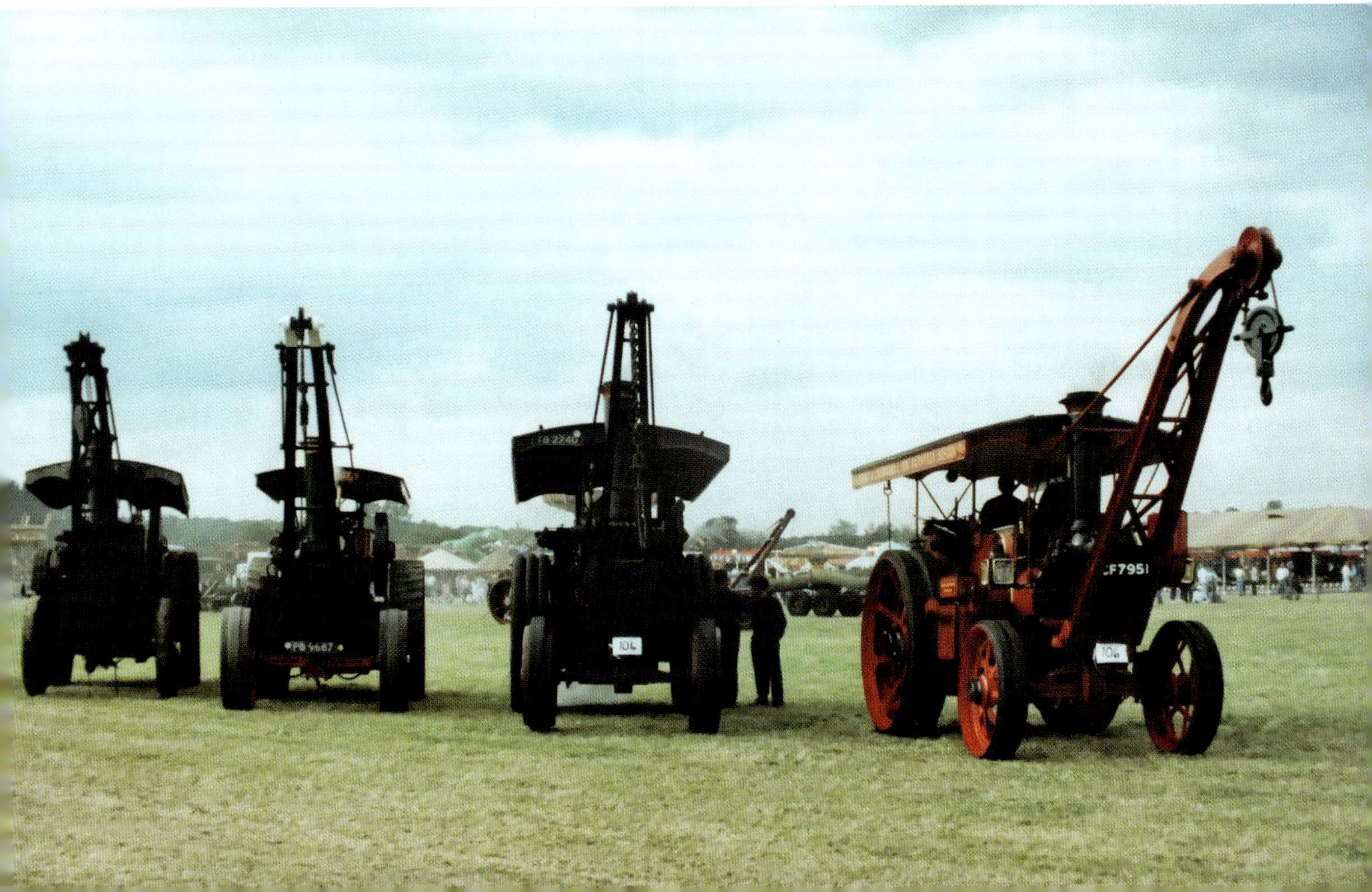
CF7951

Above: In 1994, Burrell ploughing engine No. 777 *The Earl*, dating from 1879, was in steam. This and No. 776 *The Countess* are resident at the Museum of East Anglian Life, Stowmarket, and are the only surviving ploughing engines of this make. They both worked together for the first time in preservation at the 1986 Weeting Rally. Both were once part of the famous collection of Mr Paisley from Holywell.

Opposite above: In 1997, Fowler road locomotive XC 9653 *The Lion* appeared in Pickford's colours. A year earlier it had been at Knowl Hill in wartime grey (see p. 51). The engine was sold in 1918 to London Traction Haulage who were taken over by Pickford's in 1921.

Opposite below: Also visiting in 1997 was this unusual steam roller from Switzerland built by SLM – *Schweizerische Lokomotiv-und Maschinenfabrik*. This was entered by the Dampfwalzer-Club Schweiz.

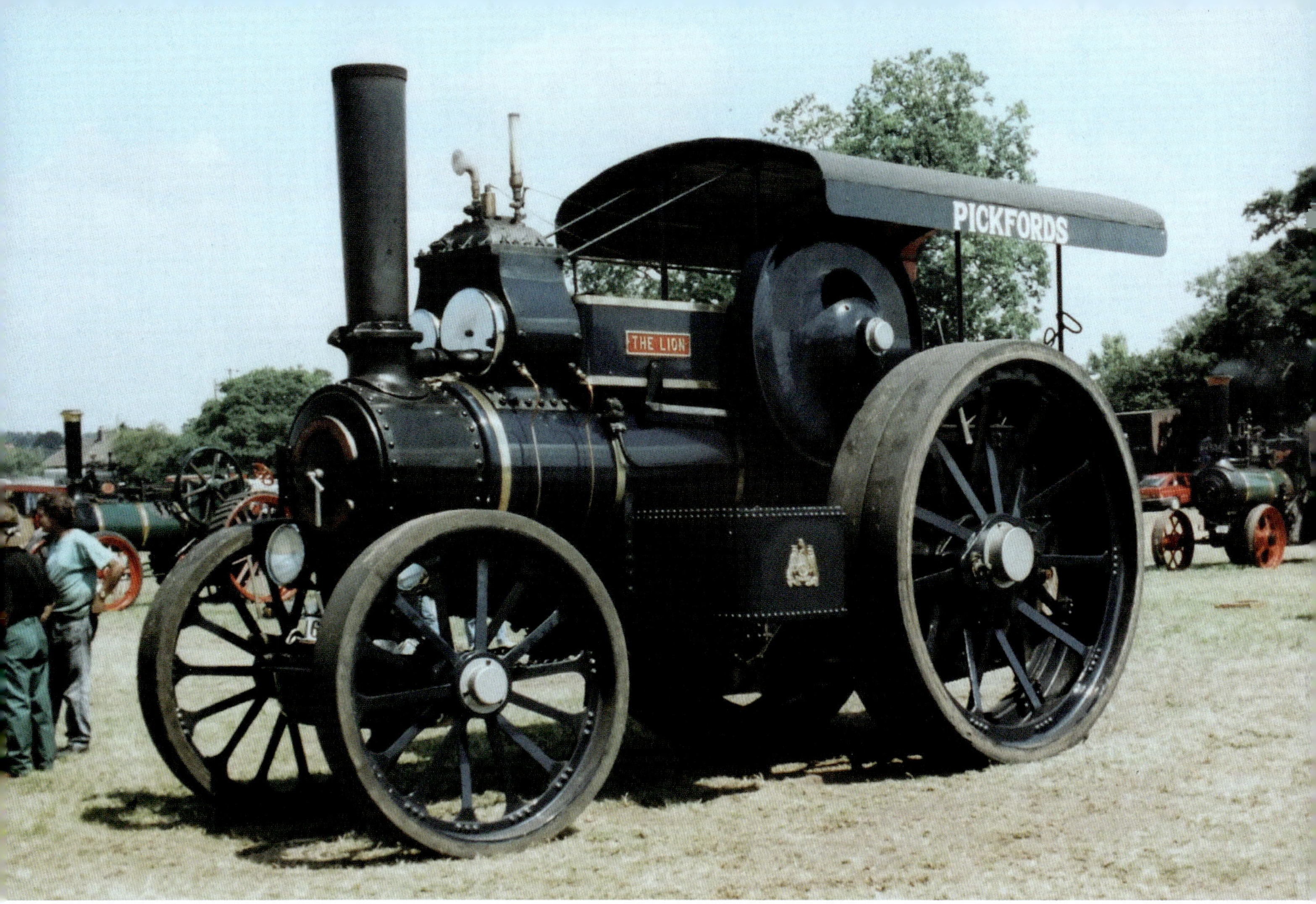
PICKFORDS
THE LION

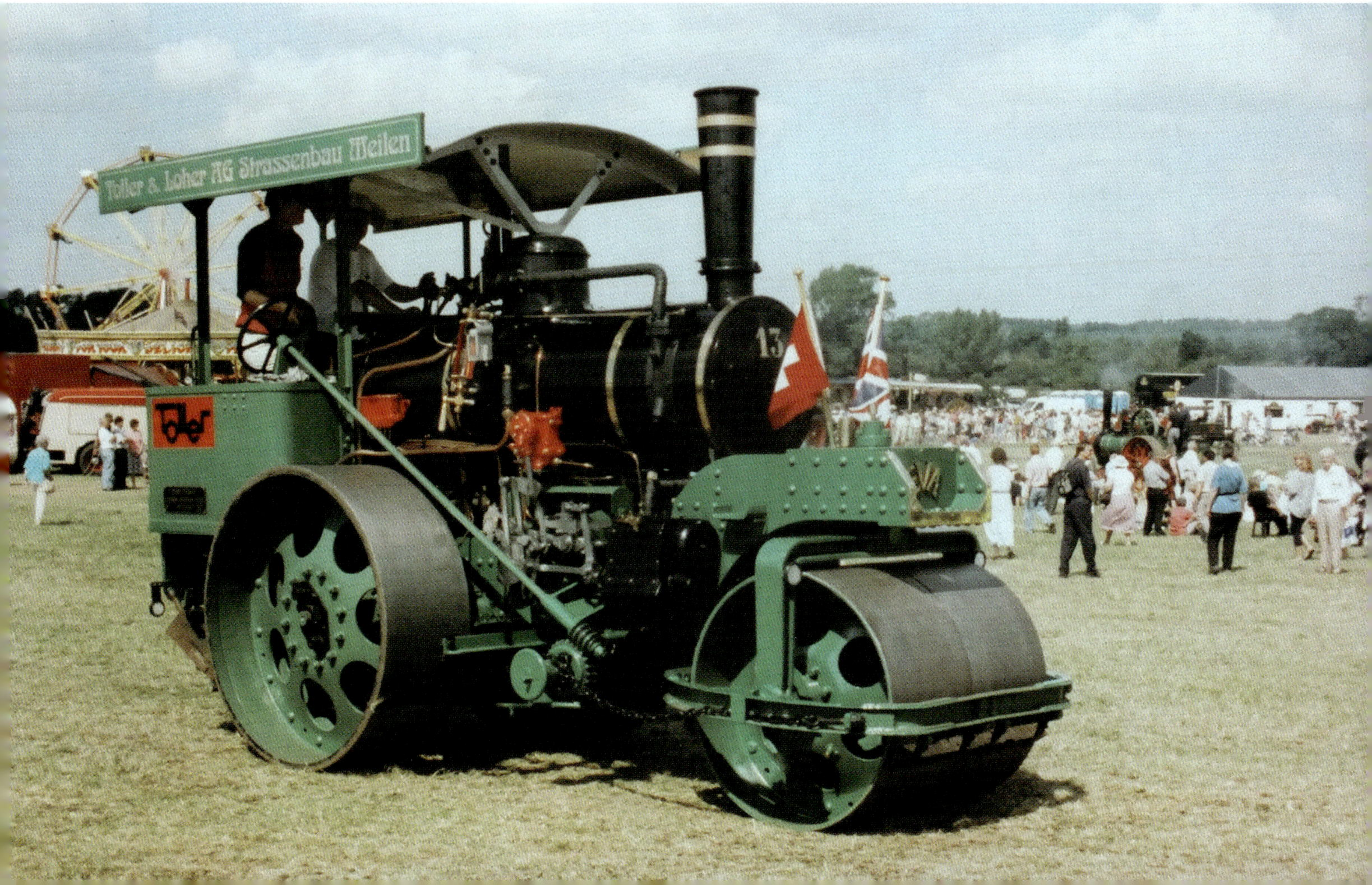
Toller & Loher AG Strassenbau Meilen
13

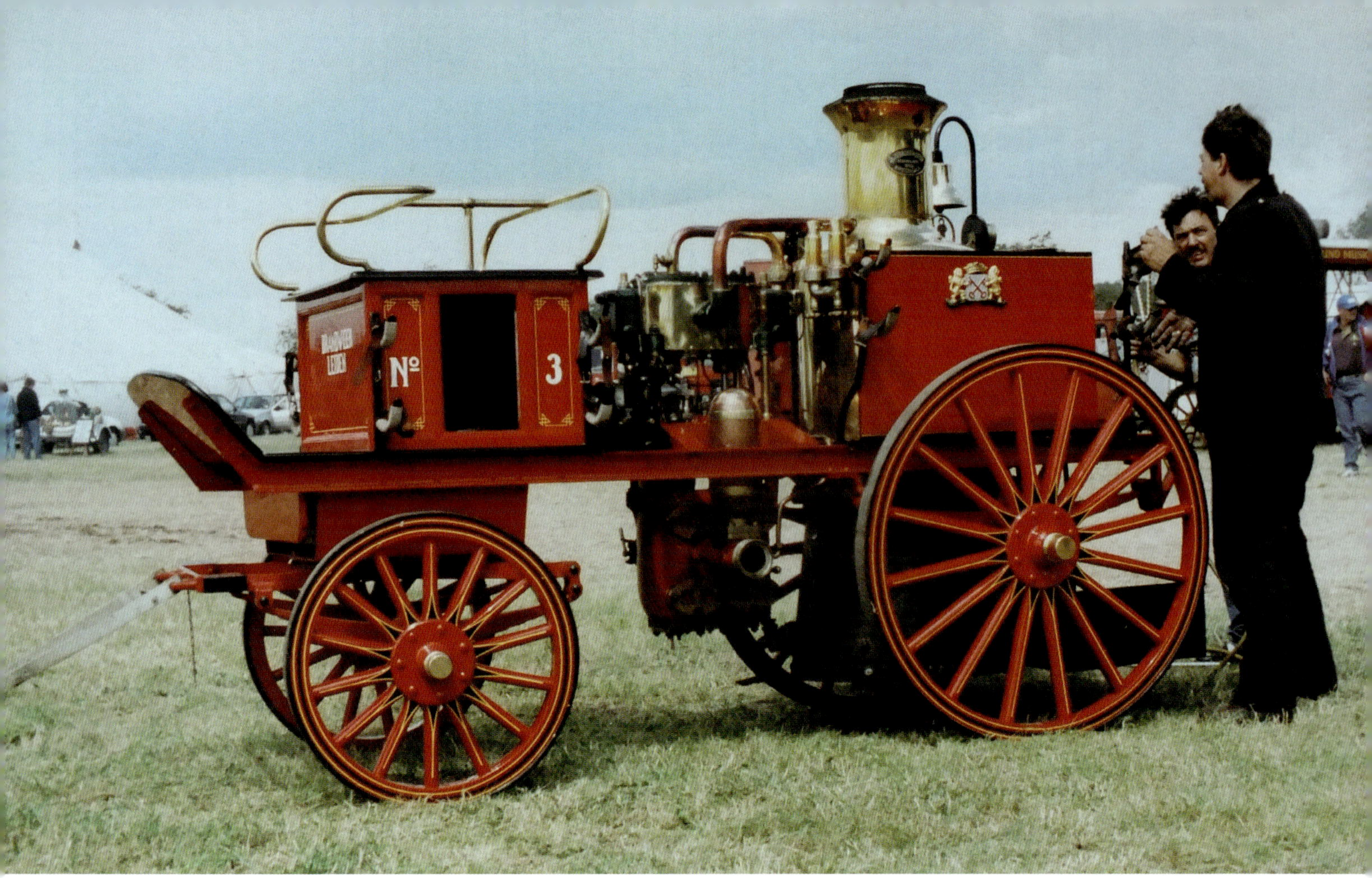
No 3

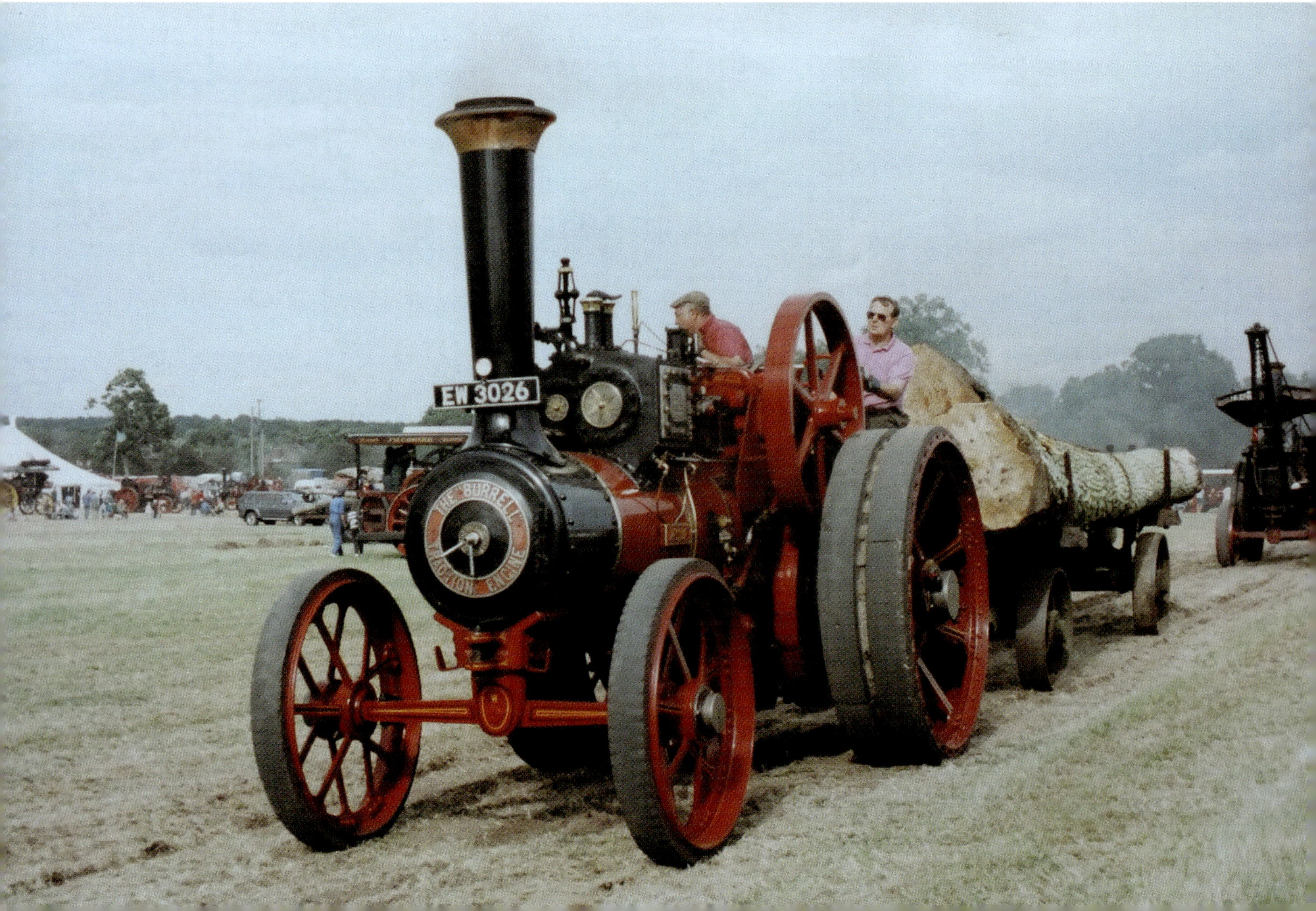
EW 3026
THE BURRELL
TRACTION ENGINE

Opposite above: At the 1998 rally, this horse-drawn steam fire engine was brought over from Holland. It was built by Bikkers of Rotterdam in 1908 and served with Leiden Fire Brigade.

Opposite below: Burrell double crank compound engine No. 3923 *Jessie* hauls a timber trailer around the ring at the 1998 rally. Built in 1922, it was exhibited new at the Royal Show at Cambridge. It was then sold to the Freeman family who have owned it ever since.

Right: Compared to other makers, Burrells made very few portables and only in earlier years. Only three are listed in the Traction Engine Register and No. 2363 is the most recent, dating from 1901. This was visiting from the Bressingham Steam Museum at Diss in 1998.

Below: A rarity is BP 9987, the only tractor surviving in the UK from the firm of Brown & May, Devizes. Dating from 1909, this is owned by Alan Rundle who also owns the Brown & May showman's road locomotive *General Buller*. This was at Weeting in 2001.

Above: The five New Zealand Burrells that were in the UK during 2001 naturally visited Weeting and the Charles Burrell Museum at nearby Thetford. This is No. 3522, an 8nhp engine built in 1913, which was now owned by the Butterick family.

Opposite above: Only three engines were built by the small firm of John M. Collings of Bacton, Norfolk. Of these, No. 2 survives. Built in 1910 with a boiler constructed by Dodman & Co., AH 047 *Bacton Hall* was at Weeting in 2003 carrying incorrect registration AH 47.

Opposite below: The main highlight at Weeting in 2008 was the first rally appearance in over thirty years of the 1920 built Burrell showman's road locomotive No. 3833 *Queen Mary*. This was again present, and in steam in 2009 when photographed here. This engine was bequeathed to the Charles Burrell Museum at Thetford by Viv Kirk, in order to secure the engine for posterity.

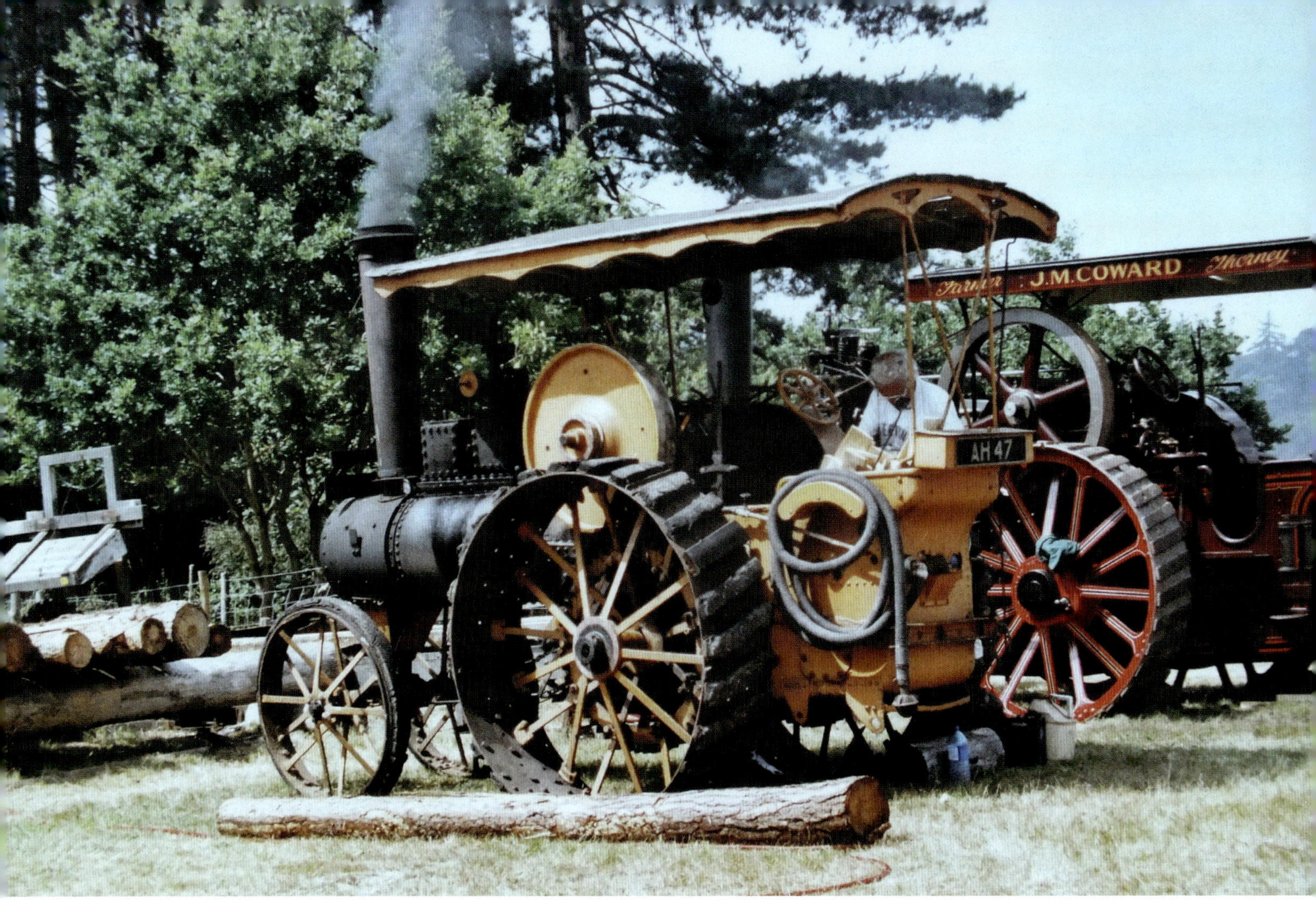
J.M.COWARD
AH 47

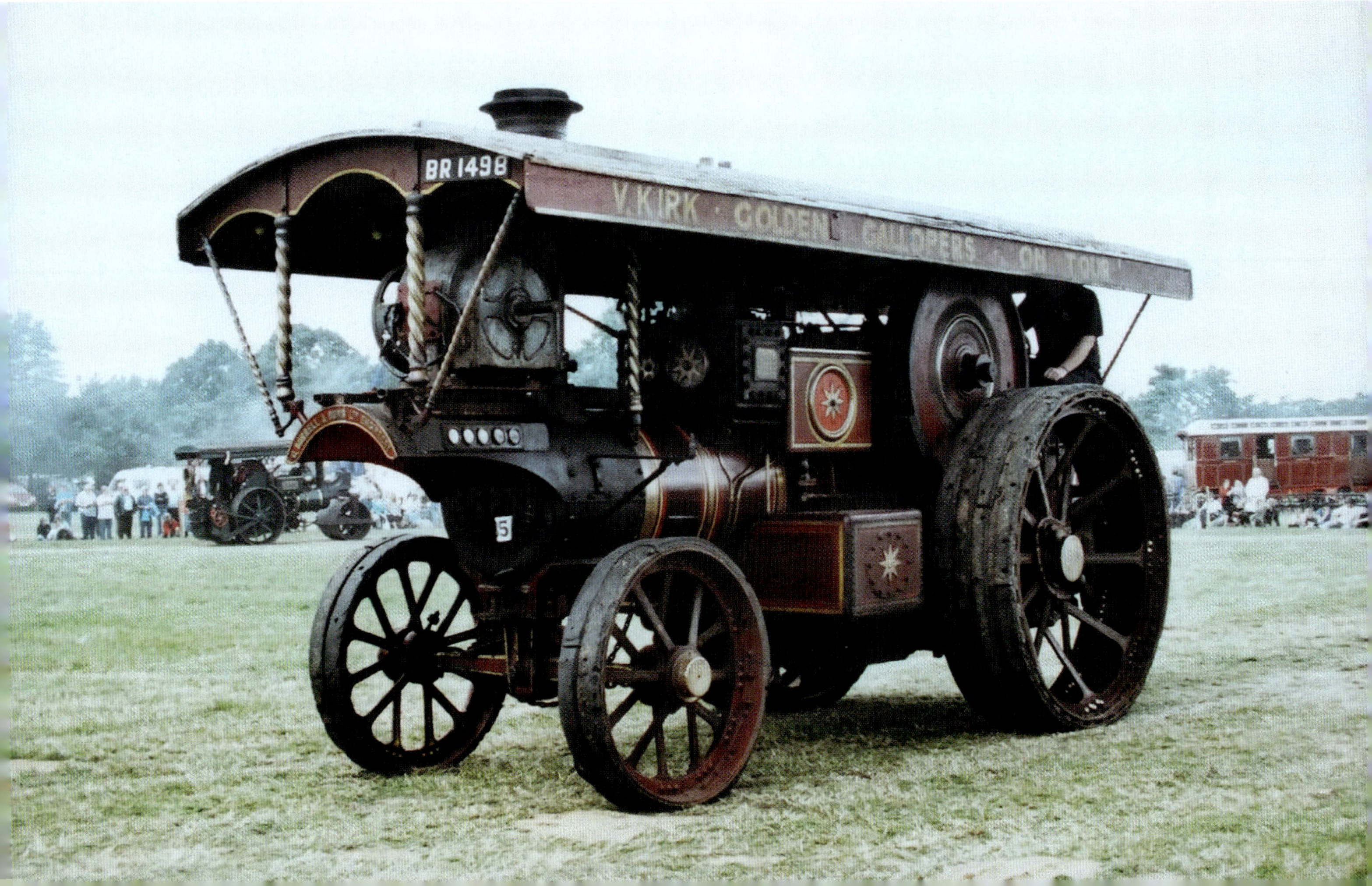
BR 1498
V. KIRK · GOLDEN · GALLOPERS · ON TOUR

Above: 2008 and Sentinel Standard wagon AW 4131 of 1918 takes on water from a hydrant in Fengate Drove, the road that leads down towards nearby Brandon. The tank on the wagon would be used to replenish engines on the rally field.

Opposite above: Engines from the Strumpshaw Steam Museum in Norfolk often attend at Weeting and two are seen together in 2008. Burrell showman's road locomotive No. 3789 *Princess Royal* was reconstructed from a traction engine in 1981. Alongside is Foden wagon No. 13802.

Opposite below: A few engine owners have had 'mini-me' miniatures constructed of their engines. At Weeting in 2009, we see Marshall BW 4509 *Old Nick* of 1908. Alongside is the half-size version built in 2001-2 and named (obviously) *Little Ol' Nick*. All measurements were taken from the full-size engine.

ANNUAL RALLIES ONGOING • 171

Left: A visit in 2018 produced a number of engines that I had not previously seen. Amongst these was Fowler tractor No. 18074 of 1930. However, the Register shows this as a reconstruction from a roller – one of more than thirty Fowler rollers to have been so treated.

Below: A feature unique to Weeting is the length of standard gauge railway track laid down in one of the fields. Various locomotives have operated along this in different years but perhaps most interesting is this Aveling & Porter geared locomotive No. 8800 of 1917 *Sir Vincent* which was present in 1994. This worked for the British Oil & Cake Mills at Erith, Kent until c. 1966, later passing to the Hollycombe Steam Collection. Another of these Aveling & Porter geared locomotives named *Sirapite* was the works locomotive at the Garratt works, and after many years in private preservation returned to the Long Shop Museum at Leiston in 2004 for restoration.

GREAT DORSET STEAM FAIR 1969–DATE (NTET AUTHORISED EVENT)

The largest and most important rally of the annual calendar is the Great Dorset Steam Fair. The origins of this date back to 1967 when Michael Oliver and a group of friends met at the Royal Oak public house in Oakford Fitzpaine to show cine films of the recently closed Somerset & Dorset line. From this, on 20 January 1968 the Dorset Steam & Historical Vehicle Club was formed and the idea of holding a Steam Party in aid of local charities was soon discussed. The first event took place on some fields near Stourpaine in 1969. The idea from the outset was to display the engines at work and the event was originally called 'The Great Working of Steam Engines'. Michael Oliver explained its origins thus 'I'd been to one or two steam engine events and I always noticed that when they took off to go home, there was more interest in the engines loading on the transporters than there was when they were actually performing in the ring'. Located on a

permanent farm site at Stourpaine Bushes from 1971, the corn would be threshed and baled, and the fields ploughed by steam, as well as by horses and tractors for comparison. Other engines would be sawing, the showman's engines would be driving rides and organs where possible and there was even space at the 400-acre site for some heavy haulage demonstrations. The GDSF proved so successful that it outgrew this site and moved to the present 600-acre Tarrant Hinton location near Blandford Forum in 1989. The event grew from a weekend event to now being held as a five-day event, attracting over 100,000 visitors each year, and organised by Michael Oliver's son Martin. Originally held in September, the month changed to August after several years of bad weather, and since 2016 it has been brought forward a week to include the Bank Holiday weekend.

A modern feature of the GDSF has been to have a special theme for each year – perhaps featuring a particular maker or type of engine. In 1995, the theme was portable engines, in 2000 Burrell engines.

ANNUAL RALLIES ONGOING • 173

For 2004, the fiftieth anniversary of the NTET was marked with a display of rarely seen engines.

This book should logically have concluded with undoubtedly the greatest rally of them all – the GDSF fiftieth anniversary event in 2018 when a total of 520 full-size steam engines was at the show including portables and unrestored engines. This led to a Guinness World Record of 472 steam vehicles which met their qualifying guidelines. There were new visitors (and some repeat visitors) from New Zealand amongst the many overseas entrants. A novel feature was a Scrap Yard area with a selection of unrestored engines showing the state from which many of the other exhibits had been rescued.

Unfortunately, I didn't get there! I had set out with a friend on the fully laden 10.00am special bus from Salisbury on the Saturday. We got as far as the Handley roundabout but then ran into a massive traffic jam. On previous visits in the 1990s when I had hit a queue here, I had arrived on site before 1.00pm. But this queue wasn't continuously slow moving, it was static for 5-10 minutes then would move forward maybe 100 yards and repeat. By 1.00pm we had only advanced about two miles. We finally arrived at the GDSF at 4.30pm, six and a half hours after departure and at an average speed of 1 mile an hour for the last six miles. We had intended to come back on the 5.15pm bus. As we could not be sure when, or even if there would be this or a later bus back to Salisbury, we decided to abandon the event and return on the bus we had arrived on. I later heard that the GDSF organisers had brought in new contractors in 2018 to organise traffic control (and toilets) for the expected additional demand that year. Unfortunately, these contractors failed to deliver properly resulting in the subsequent chaos.

So instead, this book concludes with the 2019 GDSF, when thankfully there were no transport problems. The theme was 'City of Lincoln Engines' featuring the engines and other agricultural products of five companies who had their works in the City. There was the First World War display retained from its installation between 2014-18, and also a Second World War display marking the eightieth anniversary of the start of the war. The 2020 and 2021 rallies were cancelled because of the Coronavirus pandemic, but the GDSF was back in 2022.

Below: A Dorset based roller in Dorset. This is David Antell's Wallis & Steevens 'Advance' roller No.8100 of 1936. In their working days, rollers often towed living vans from site to site for the crews that worked them. This living van, however, appears to be a former railway brake van mounted on road wheels. Taken at Stourpaine in 1986, my first visit to this event.

Above: Being sited on farmland, the Great Dorset has been able to offer ploughing demonstrations at both sites. At Stourpaine in 1987, a single-ended plough is being pulled across by the engine out of sight behind the photographer.

Below: Of the surviving engines from the Devizes based company of Brown & May Ltd, all are portable engines except one tractor and one showman's road locomotive. I first encountered the latter, No. 8742 of 1912 *General Buller*, at Stourpaine in 1987.

Above: 1988 was the last year at Stourpaine and from 1989 the GDSF moved to its current permanent site at Tarrant Hinton. This is 1947 Aveling-Barford roller No. AH162. Aveling & Porter became Aveling-Barford from 1932 but did not adopt the new name and makers number series until 1937. Production switched from Rochester, Kent to Grantham, Lincs, with steam roller production ending c. 1950.

Opposite above: Heavy haulage, 1988 style! There was no 'playpen' in place then and loads were hauled through the public areas. What would Health & Safety say? Here we have 1914 Burrell road locomotive KE 3865 *Duke of Kent* leading, followed by BE 7217 *Lord Roberts* and McLaren WF 1864 *Boadicea*.

Opposite below: Rear view of the same load. The 'Amalgamated Heavy Haulage' machinery trailer came from the Central Electricity Generating Board.

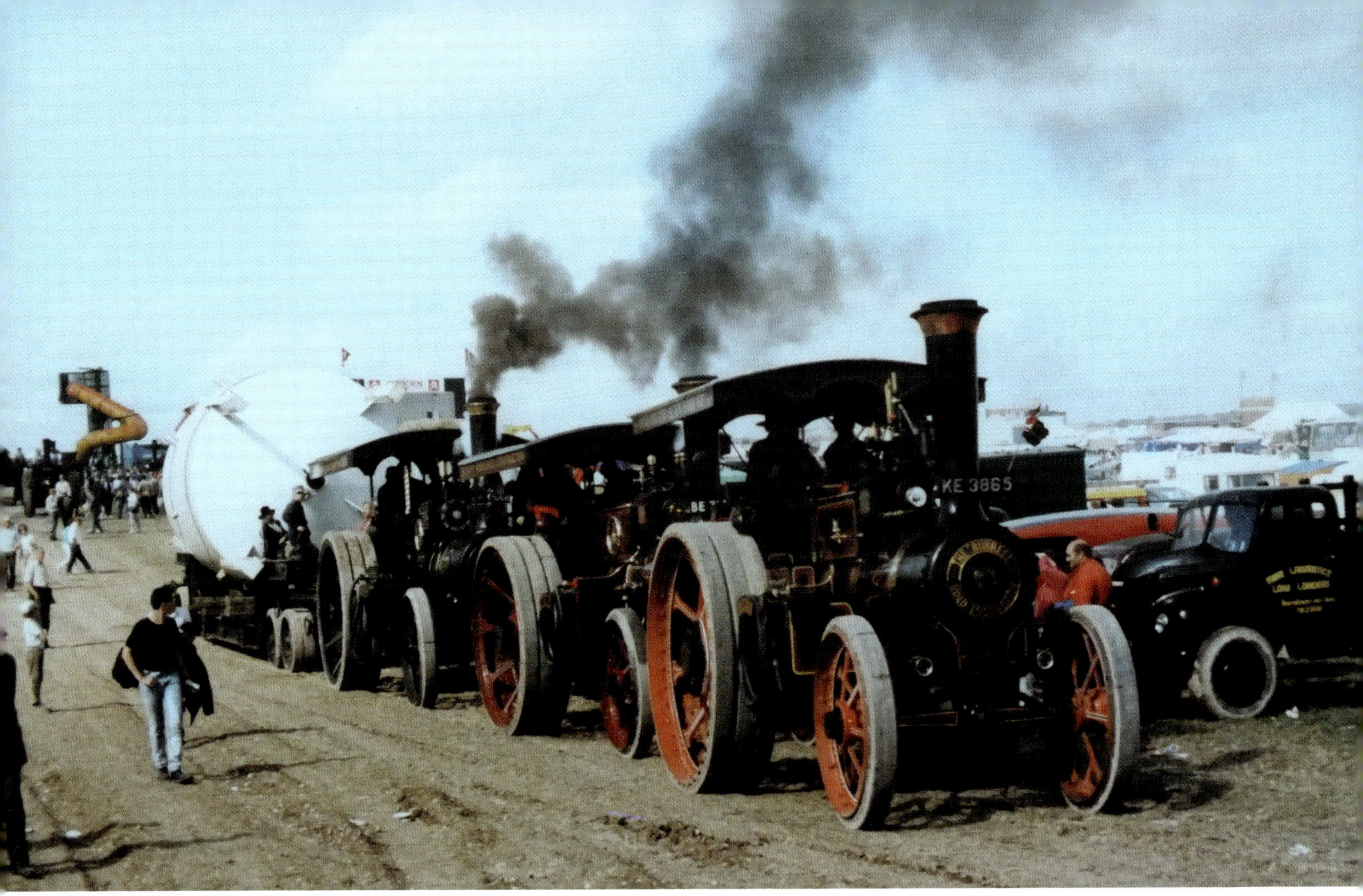
KE 3865

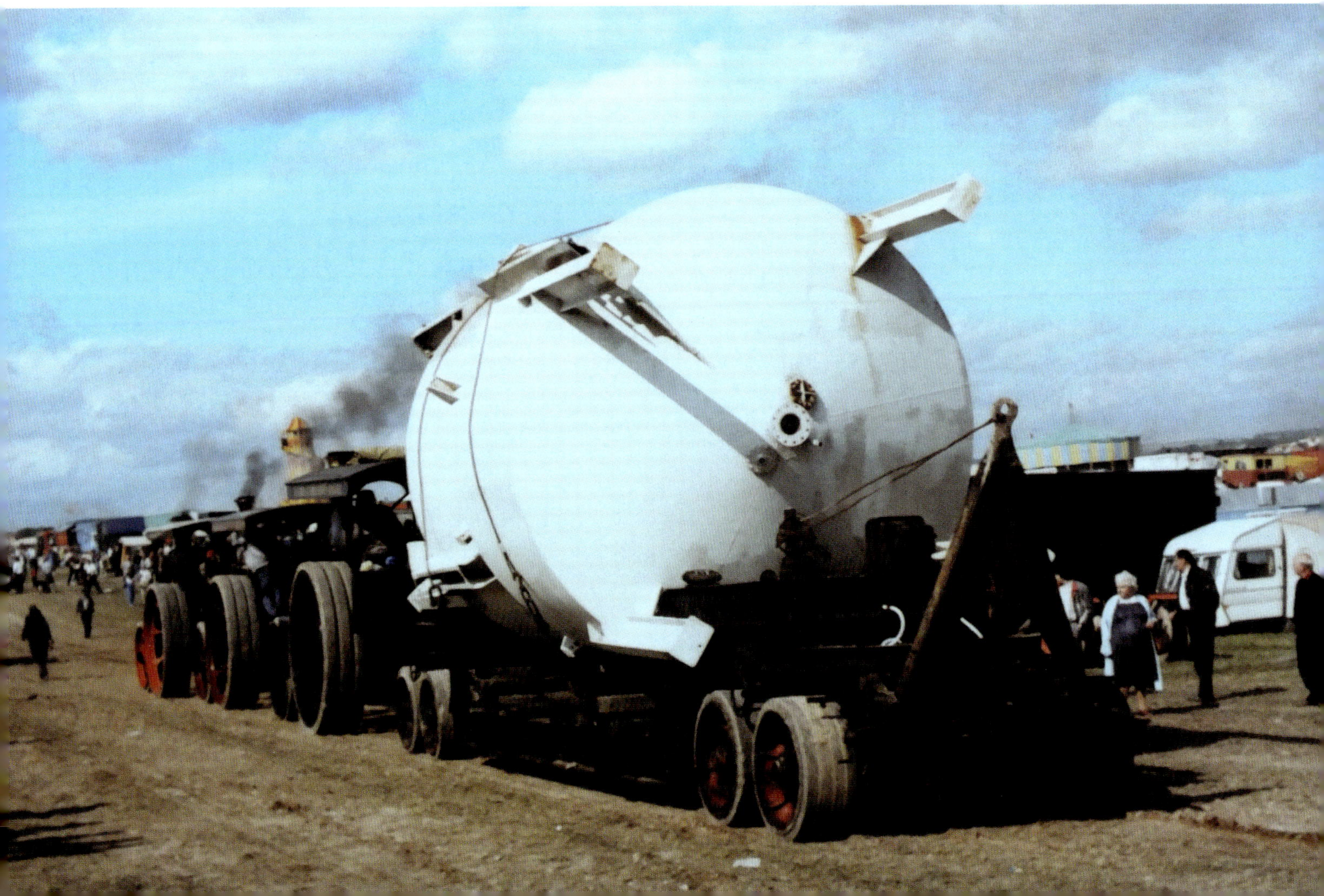

Above: This odd-looking roller is Marshall No. 87125 of 1933. A vertical boiler tandem roller, it is the only survivor of its type. This was an entrant in 1990 at Tarrant Hinton.

Opposite above: A highlight of the GDSF is the rows of showman's engines lined up by the fair and powering some of the rides. The fair continues into the night and then the engines are lit up with hundreds of bulbs under their canopies. The showmen elaborately decorated their engines with twisted brass fittings and bright paintwork, but here are a pair of engines somewhat plainer but just as imposing. Fowler Nos. 15652-3, *Repulse* and *Renown* were built for County Durham showman John Murphy in 1920 to work his 'Proud Peacocks' ride. *Renown*, on the right has written on the front of the canopy 'Plain but powerful'. *Renown* was also fitted with a crane for erecting and dismantling the ride. They were photographed together in 1993.

Opposite below: Ploughing in 1993 took place on a field at a higher elevation than the rest of the Fair which is therefore out of sight below. In action is Fowler K7 engine KE 3135 *General French*.

STEAM
COLUMBIA
Mighty in War & Peace
JOHN MURPHYS MODERN AMUSEMENTS ON TOUR
JOHN MURPHYS PROUD
31
CU 977

KE 3135

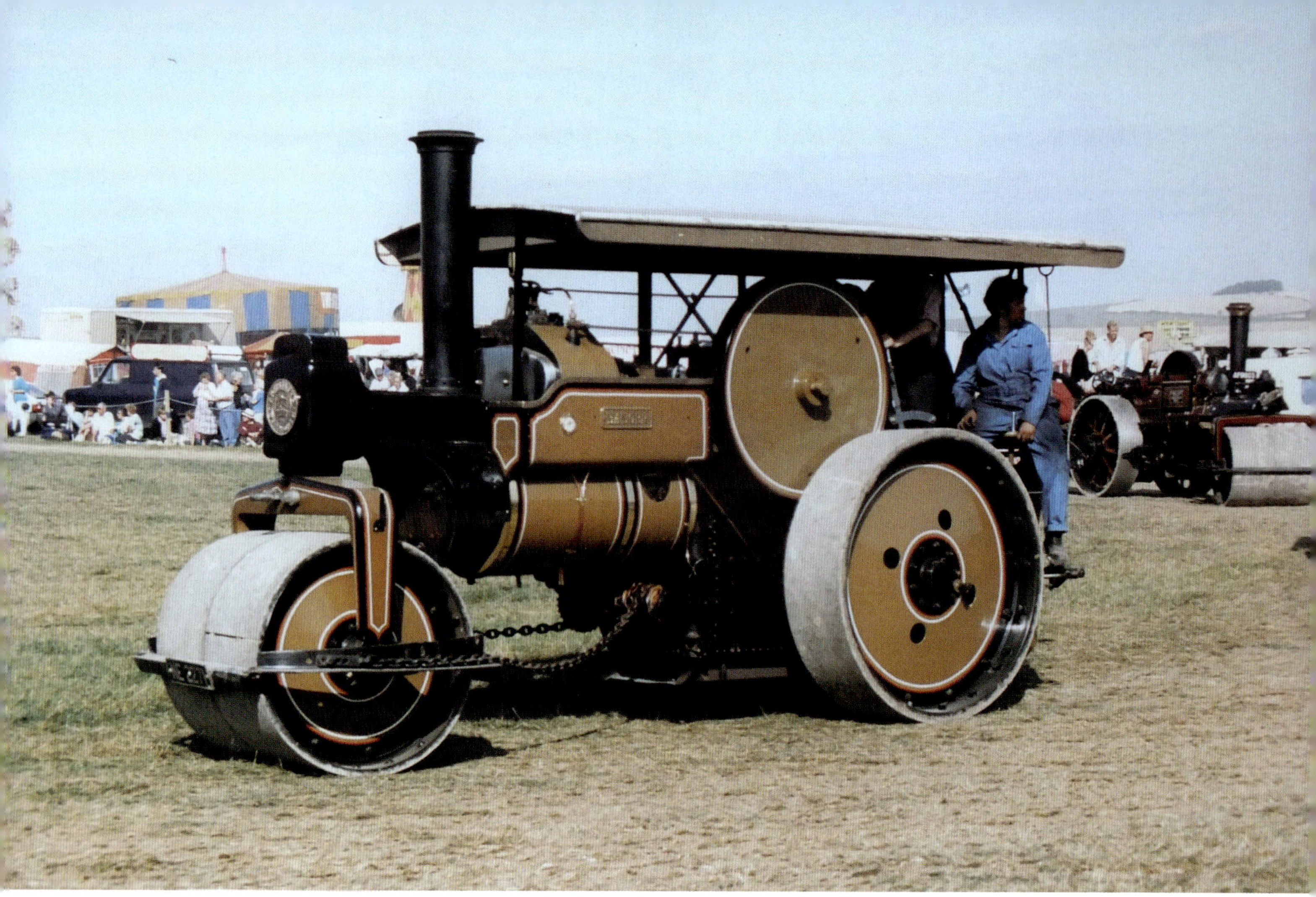

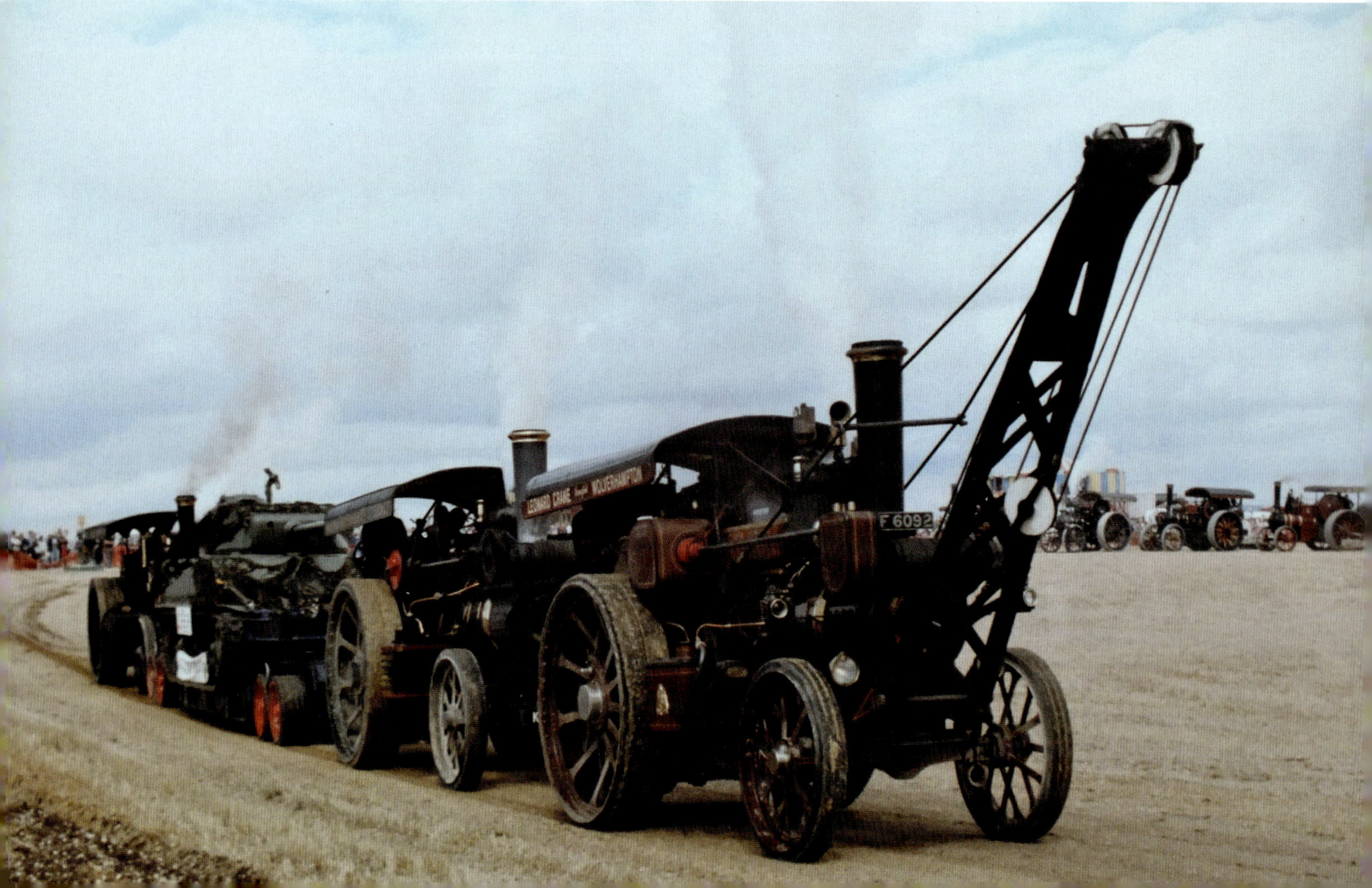
F 6092

Opposite above: Sir W.G. Armstrong-Whitworth Ltd made a number of handsome looking rollers in the 1920s. All seven surviving engines are compounds with piston valves. Circulating in one of the side rings at the 1993 show is NE 2771 *Badger* of 1925.

Opposite below: What the GDSF is most noted for is its haulage displays. A variety of different loads are assembled each year for the haulage boys to play with. In 1995, a Sherman tank was brought in, hauled by steam from the Bovington Tank Museum in Dorset. Taking it around the 'play pen' we have Fowler crane engine *Wolverhampton Wanderer* on the front. Behind is Fowler KD 2846 *Duke of York* (with jib removed). Another road locomotive is on the rear acting as brake engine.

Above left: A special feature for the 1995 show was portable engines, and the chimneys of several of these can be seen in this picture. The engine depicted is by Barrows & Stewart, No. 1098 dating from c. 1870.

Above right: There were portables by many makes and this is a French made engine from Brouhot et Cie, Vierzon-Ville, Cher. No 5215 is now kept at the Bredgar & Wormshill Railway near Maidstone, Kent.

Above: From 1996, this is Fowler B6 road locomotive No. 17105 *Atlas*. It was painted in the livery of its original owners Norman E. Box Ltd and paired with a matching low loader trailer. This company was taken over by Pickford's in 1930.

Opposite above: Foden wagon No. 11340 was originally a dray wagon for a Brighton brewery but in the 1960s was fitted with a replica body to that once carried by the Foden Motor Works Band bus 'Puffing Billy'. After periods with owners from Norfolk and County Durham, M 6359 is now with the Searle family collection from Horsham and is a regular attender at this and other events, as here in 1998.

Opposite below: There are always several working engines in the threshing and sawing areas. From 2003, this is Marshall 6nhp engine No. 86265 of 1931. In the distance can be seen sticking up some of the modern fairground rides – just part of the attractions on this vast 600-acre site.

TA1935
M6359
HORSHAM
TRACTION
CO
TEL
195
19
STEAM OMNIBUS' CO.

OD 177
GRANDE OLDE

Above: In 2015, Marshall No. 73901 tows one of the trailers that frequently give rides round the 'play pen', raising money for charity from the fares. That front number plate seems rather too small!

Below: 2016 was notable for the completion of not just one but both of the surviving Foden undertype wagons. For years, Foden had persisted with overtype wagons with their in-line boilers and cylinders atop, traction engine style. This all took up valuable load space. They developed the undertype wagons with vertical boiler, as built by rivals Sentinel in a last-ditch attempt to promote the steam wagon. But it was too late, the diesel lorry had arrived, and legislation discriminated against steam. After 1933, Foden would switch entirely to making diesel lorries and buses. LG 4815 is a Speed 6 model of 1930, while MI 3304 is a 1931 Speed 12, which originally worked in Ireland.

Above: The heavy haulage displays have been enhanced by newly restored and imported engines. Here Foden No. 2701 *Tombola,* an 8nhp engine that came back from Tasmania in 1994, leads *Colossus*, the McLaren tandem cylinder 10nhp road locomotive No. 897 that was brought back from Argentina (see also p. 36). The trailer with the transformer load is the same one that was hauled from London to Brighton in 2011 (see p. 96). Seen in 2016.

Right: Each year there are some engines displayed inside marquees from organisations like the NTET or the Steam Plough Club. In 2016 newly restored Burrell road locomotive No. 3804 *Independence* was a static exhibit. It would back next year in steam.

Above: This Tasker's 'Little Giant' tractor, entered in 2016, is unusually fitted with pannier tanks either side of the boiler, rather in the style of a Great Western Railway locomotive.

Below: There is always something new to see at the GDSF and one of the 2017 entrants that was new for me was this Garrett wagon. SX 3925 had been in Canada until 2012 and is the only 6-ton overtype superheated steam wagon left. Both this and the Foden Speed 6 wagon in 2016 had been entered by owners from Scotland.

Right: Once again, the weather was excellent in 2017, so no excuses for another haulage photo even if it features some of the 'regulars'. Fowler crane engine SG 4713 *The Great North* teams up with McLaren 1912 10nhp road locomotive No. 1332 in War Department livery and an ex. Pickford's Diamond T tractor lorry to haul the massive 100-ton trailer.

Below: From 2014-2018 the nation was celebrating the centenary of the First World War and the GDSF took a major role by creating a trenches experience area. There were trenches, field workshops, artillery emplacements etc., all enhanced by vehicles of the period. Foden wagon M 8562 dating from 1916 is seen among a selection of contemporary lorries in 2017. This wagon originally saw service in France during the First World War being used on repairs to roads following shelling damage, After the war it was sold to a French company from Cambrai where it was used until 1948. The wagon is fitted with buffers either side of the smokebox for shunting railway wagons.

Left: The theme for 2019 was engines made by the Lincoln based companies of Clayton & Shuttleworth, Foster, Ruston Proctor, Ruston & Hornsby and Robey. The two earliest surviving traction engines from William Foster & Co. are seen together. On the left DO 1945 *Master Fred* is No. 2163 of 1896, while DO 2003 *Jumbo the 4th* is No. 2127. The Foster marquee can be seen in the background.

Below: Inside the Robey display marquee is wagon No. 42657 of 1925. This was new to Highways Colloidal Ltd and has been with Robert Crawford of Boston since 1966.

Above: Robey were noted for their tandem rollers. This is No. 42129 *Bullet* of 1924. As with other tandem rollers and the two surviving wagons, it is fitted with Robey's stayless circular 'thimble' firebox.

Right: In the sawing area, a detailed close-up of the cylinders of a Sawyer-Massey traction engine, which, as the plate shows was made in Canada. Built in 1912, this was retired in the 1970s and moved into a museum in 2008. A return to steam was made in 2018 and the engine was visiting the GDSF from Canada.

Above: We saw the one-off Foreman replica of a Wallis & Steevens tractor before on the Isle of Wight (see p. 142) This is how it looks now, owned by Andy Selwood since 1994 and now with a canopy and named *The Hamster*. It can usually be found in the working area and 2019 was no exception.

Left: One of the engines painted in wartime livery for the First World War centenary was Aveling & Porter roller No. 7857 of 1912.

Right: Each year there are usually a few entries from Ireland and in 2019 one of these was McLaren No. 1497 a 1917 built road locomotive. Although now coming from Carlow, this was not originally sold to an Irish customer but was repatriated from South America in 2003.

Below: Plenty of activity going on in the 'play pen' where there is always something on the move, and it is not just the big boys that come out to play! Although this is the main area, in recent years a section has been set up to one side where the smaller miniatures can also perform in safety while the public can enjoy watching the performance.

BIBLIOGRAPHY

COLBECK, Simon (ed.), *Old Glory Archive vol 5: the colour files*, Yalding, Kelsey Publishing, 2021
EDMONDS, Chris, 'A wager for ale' – extracts in the Woodcote Rally programme 2010
FINCH, Barry J., *A rally of traction engines*, Shepperton, Ian Allan, 1969
JOHNSON, Brian, *The Traction Engine Register, 13th edition*, Horsham, Southern Counties Historic Vehicles Preservation Trust, 2020 and earlier editions
JOHNSON, Brian, *Steam traction engines, wagons and rollers in colour* Poole, Blandford Press, 1976
LOCKETT, David & ARLETT, Mike, *Traction engines: a colour portfolio*, Horsham, Ian Allan, 2002
MILLS, Kevin (ed.), *60 years of the National Traction Engine Trust*, Redditch, NTET, 2014
SAWFORD, Eric, *The heyday of the traction engine*, Shepperton, Ian Allan, 1995
SAWFORD, Eric, *Steam Power, Farm & Highway*, Burton-upon-Trent, Trent Valley, 1988
SAWFORD, Eric, *Steam wagons in colour*, Shepperton, Ian Allan, 1997

MAGAZINES AND PERIODICALS

Old Glory (Yalding, Kelsey Media) Monthly magazine published since 1988
Steaming (Redditch, NTET) Quarterly magazine of the National Traction Engine Trust
Vintage Spirit (Cranleigh, Steam Heritage Publishing) Monthly magazine published since August 2002
Rally programmes from events I have visited